Handbook of Emergency Management Concepts

A Step-by-Step Approach

Handbook of Emergency Management Concepts

A Step-by-Step Approach

Michael L. Madigan

CRC Press
Taylor & Francis Group
Boca Raton London New York

CRC Press is an imprint of the
Taylor & Francis Group, an **informa** business

CRC Press
Taylor & Francis Group
6000 Broken Sound Parkway NW, Suite 300
Boca Raton, FL 33487-2742

Printed on acid-free paper

International Standard Book Number-13: 978-1-138-56853-2 (Hardback)

Library of Congress Cataloging-in-Publication Data

Name: Madigan, Michael L., author.
Title: Handbook of emergency management concepts : a step-by-step approach / Michael L. Madigan.
Description: Boca Raton, FL : CRC Press, 2018. | Includes bibliographical references and index.
Identifiers: LCCN 2017031852| ISBN 9781138568532 (hardback : alk. paper) | ISBN 9780203704257 (ebook)
Subjects: LCSH: Emergency management.
Classification: LCC HV551.2 .M3265 2018 | DDC 363.34--dc23
LC record available at https://lccn.loc.gov/2017031852

Visit the Taylor & Francis Web site at
http://www.taylorandfrancis.com

and the CRC Press Web site at
http://www.crcpress.com

To my beautiful wife Michelle, who has sharpened my wit, life, and soul that have evolved our life together. Who think the same thoughts without need of speech, and babble the same speech without need of meaning. This why I love you and hold you dear and precious in my heart.

My best friend and battle comrade, SFC Joseph F. Johnston, who has been there through thick and thin times.

To my sons, Michael Madigan and Mark Walker Madigan, and my daughters, Kristen Genco, Nikole Madigan, Leisha Eshbach, Angela Bryant, Neitha Engert, and Rachel Walker. I want to give a huge thanks to my mentor and friend Daniel Arden, who has always been there with advice, direction, and leadership throughout my military and civilian career. I especially want to thank all of the Dragon soldiers of the U.S. Army Chemical Corps and the instructors and cadre of the Incident Response Training Department. HOOAH

Contents

Preface

This book has been prepared for anyone who wants to know about emergency management and those professionals whose responsibility is to protect the public and is based on the most reliable hazard awareness and emergency education information available at the time of publication, including advances in scientific knowledge, more accurate technical language, and the latest physical research on what happens in disasters.

This publication cannot cover every factor, situation, or difference, as no plan can really do so, but covers preplanning and mitigation efforts to help alleviate the disruption of the incident/disaster in buildings, infrastructure, or other environmental features that might be of interest.

This book can be used as a reference source or as a step-by-step process. The focus of the content is on how to develop, practice, and maintain emergency plans that reflect what must be done before, during, and after a disaster to protect people and their property.

The Planning process and the preparations for all types of disasters will ensure that the committees planners will complete the preplans and mitigate as much as possible that are recognize within their communities prior to an incident which will allow the communities to be better prepared for event of any unlikely emergency scenarios.

Information is provided on how the public can be ready in case of a national emergency—including a possible terrorist attack involving biological, chemical, or radiological weapons. Another source for the general public can be found by logging on to the Department of Homeland Security's website, http://www.ready.gov, or by calling 1-800-BE-READY for printed information.

All of the recent devastation caused by tsunamis, hurricanes, and wildfires highlights the need for highly trained professionals who can develop effective strategies in response to these disasters. This invaluable book will assist those with the tools to address all phases of emergency management. It covers everything from the social and environmental processes that generate hazards to vulnerability analysis, hazard mitigation, emergency response, and disaster recovery.

Author

Michael L. Madigan works as a master instructor for the Department of Homeland Security for the Civil Support Teams and Hazardous material emergency response operations on weapons of mass destruction (WMD), installation emergency management, and threat assessments to the United States Army Chemical, Biological, Radiological and Nuclear School (USACBRN) on Fort Leonard Wood in Missouri. He is also retired from the United States Army and served as a senior chemical operation specialist to the USACBRN School, Central Command (CENTCOM), and the Multi-National Forces Command (MNFI) in Iraq.

Mr. Madigan is married to his wife Michelle and has 7 children and 14 grandchildren. He loves to travel.

He earned a BA degree in Criminal Justice from the University of Massachusetts and a Master of Security Management from the American Military University from Charles Town. In addition, he holds the Certificate in Security Management from the University of Massachusetts, and is currently working on his doctoral degree in Emergency Management (DEM) from Capella University. Mr. Madigan is also a certified fire instructor Level II, hazardous material technician train the trainer, incident command system train the trainer, installation emergency management instructor, WMD/terrorism response instructor, certified Incident Command System/National Incident Management System (ICS/NIMS) instructor, and advanced chemical biological instructor.

Over his 40-plus-year career, he has distinguished himself by exceptionally meritorious service in a succession of positions of great importance and responsibility to the Army and the Nation.

His previous positions directly influence the Operations for the J3 MNFI CBRNE, United States Forces Iraq, and Iraqi Army WMD/Terrorism/CBRNE Non-Commissioned Officer in Charge (NCOIC) Team Chief, and Survey team chief of a civil support team.

As the WMD/Terrorism and CBRNE Team Chief, he has positively and directly influenced the Army and the Department of Defense Policies and program. He has made recommendations to the MNFI Commander, and has communicated advice and recommendations to reflect the depth and breadth with considerable experience and affected future programs and policy that will carry the Army through the end of the sustained combat operations and into an Army focused on building the capacity of the partners of the United States dealing with CBRNE and hazardous materials research, training, and response.

Mr. Madigan is also an adjunct professor for homeland security-related topics, emergency management, mass casualty/mass fatality, terror/terrorism history, related operations, and WMD at Central Texas College on Fort Leonard Wood Campus, Missouri.

Introduction

There are many ways to describe emergency management and the importance of the tasks that emergency managers perform. Indeed, in some respects, it hardly seems necessary to explain the need for a profession whose purpose is saving lives and property in disasters. It is likely that, while many people recognize that their communities are exposed to environmental threats requiring a systematic program of protection, only a few appreciate the magnitude and diversity of the threats. One can introduce the study of emergency management by noting that losses from disasters—in the United States and the rest of the world—have been growing over the years and are likely to continue to grow.

The world is plagued by an increasing number and variety of types of disasters, an impression that is certainly heightened by what seem to be frequent, very-large-scale natural disasters—including earthquakes, floods, hurricanes, volcanic eruptions, and wildfires—all over the globe. When we add to these events a wide range of severe storms, mudslides, lightning strikes, tornadoes, and other hazard agents affecting smaller numbers of people, one might conclude that natural disasters are increasing.

Technological activities also initiate disasters. Hazardous materials are transported via road, rail, water, and air. When containment is breached, casualties, property loss, and environmental damage can all occur. Some technologies, such as nuclear power plants, pose seemingly exotic risks, whereas more commonplace technological processes such as metal plating operations use chemical agents that are very dangerous. Even the queen of American technology, the space program, has experienced disaster associated with system failures. Finally, we see terrorists operating on U.S. soil—made forever visible by the attacks on the World Trade Center on September 11, 2001.

With increasing population density and changing land use patterns, more people are exposed to natural hazards, and consequently, our accumulated human and economic losses are increasing. Much of this exposure is a matter of choice. Sometimes, people choose hazardous places, building houses on picturesque cliffs, on mountain slopes, in floodplains, near beautiful volcanoes, or along seismic faults. Sometimes, people choose hazardous building materials that fail under extreme environmental stresses—for example, unreinforced masonry construction in seismically active areas. Some exposure results from constrained choices; the cheap land or low rent in flood plains often attracts the poor.

Given the increasing toll from disasters arising from natural hazards, technological accidents, and terrorist attacks using technological agents, American society must decide whether the risks are "acceptable." Moreover, given the limited amount of time and resources that can be devoted to risk management, decisions must be made about which risks to address.

When individuals, organizations, or political jurisdictions reach consensus that a given risk is unacceptable, resources can be marshaled to reduce the risk to some level deemed more acceptable. Such resources can be used to attempt to eliminate the source of the danger or, alternatively, change the way people relate to the source of danger.

In general terms, emergency management is "the discipline and profession of applying science, technology, planning and management to deal with extreme events that can injure or kill large numbers of people, do extensive damage to property, and disrupt community life" (Questions on the emergence of a discipline/field of study training). Thus, emergency managers identify, anticipate, and respond to the risks of catastrophic events in order to reduce to more acceptable levels the probability of their occurrence or the magnitude and duration of their social impacts.

1 Emergency Management

Disaster management (or *emergency management*) is the creation of plans through which communities reduce vulnerability to hazards and cope with disasters. Disaster management does not avert or eliminate the threats; instead, it focuses on creating plans to decrease the effect of disasters. Failure to create a plan could lead to human mortality, lost revenue, and damage to assets. Currently, in the United States, 60% of businesses do not have emergency management plans. Events covered by disaster management include acts of terrorism, industrial sabotage, fire, natural disasters (such as earthquakes, hurricanes, etc.), public disorder, industrial accidents, and communication failures.

EMERGENCY PLANNING IDEALS

If possible, emergency planning should aim to prevent emergencies from occurring and, failing that, should develop a good action plan to mitigate the results and effects of any emergencies. As time goes on, and more data become available, usually through the study of emergencies as they occur, a plan should evolve. The development of emergency plans is a cyclical process, common to many risk management disciplines, such as Business Continuity and Security Risk Management, as set out below:

- Recognition or identification of risks
- Ranking or evaluation of risks
    - Responding to significant risks
    - Tolerate
    - Treat
    - Transfer
    - Terminate
- Resourcing controls
- Reaction Planning
- Reporting and monitoring risk performance
- Reviewing the Risk Management framework

There are a number of guidelines and publications regarding emergency planning, published by various professional organizations such as ASIS, the Federal Emergency Management Agency (FEMA), and the Emergency Planning College. There are very few emergency-management-specific standards, and emergency management as a discipline tends to fall under business resilience standards.

In order to avoid, or reduce, significant losses to a business, emergency managers should work to identify and anticipate potential risks, hopefully to reduce their probability of occurring. In the event that an emergency does occur, managers should have a plan prepared to mitigate the effects of that emergency, as well as to ensure business continuity of critical operations postincident. It is essential for an organization to include procedures for determining whether an emergency situation has occurred and at what point an emergency management plan should be activated.

IMPLEMENTATION IDEALS

An emergency plan must be regularly maintained, in a structured and methodical manner, to ensure that it is up-to-date in the event of an emergency. Emergency managers generally follow a common process to anticipate, assess, prevent, prepare, respond, and recover from an incident.

Preincident Training and Testing

A team of emergency responders performs a training scenario involving anthrax.

Emergency management plans and procedures should include the identification of appropriately trained staff members responsible for decision making when an emergency occurs. Training plans should include internal people, contractors, and civil protection partners and should state the nature and frequency of training and testing.

Testing of a plan's effectiveness should occur regularly. In instances where several business or organizations occupy the same space, joint emergency plans, formally agreed to by all parties, should be put into place.

Communicating and Incident Assessment

Communication is one of the key issues during any emergency; preplanning of communications is critical. Miscommunication can easily result in emergency events escalating unnecessarily.

Once an emergency has been identified, a comprehensive assessment evaluating the level of impact and its financial implications should be undertaken. Following assessment, the appropriate plan or response to be activated will depend on a specific preset criteria within the emergency plan. The steps necessary should be prioritized to ensure that critical functions are operational as soon as possible.

Phases and Personal Activities

Emergency management consists of five phases: prevention, mitigation, preparedness, response, and recovery (Mission Areas, FEMA.gov, 2017).

Prevention

Prevention was recently added to the phases of emergency management. It focuses on preventing the human hazard, primarily from potential natural disasters or terrorist attacks. Preventive measures are taken on both the domestic and international levels, designed to provide permanent protection from disasters. Not all disasters, particularly natural disasters, can be prevented, but the risk of loss of life and injury can be mitigated with good evacuation plans, environmental planning, and design standards. In January 2005, a total of 167 governments adopted a 10-year global plan for natural disaster risk reduction called the Hyogo Framework (Hyogo Framework for Action (HFA), UNISDR, www.unisdr.org/we/coordinate/hfa).

Preventing or reducing the impacts of disasters on our communities is a key focus for emergency management efforts today. Prevention and mitigation also help reduce the financial costs of disaster response and recovery. Public Safety Canada is working with provincial and territorial governments and stakeholders to promote disaster prevention and mitigation using a risk-based and all-hazards approach. In 2008, federal/provincial/territorial ministers endorsed a National Disaster Mitigation Strategy (Emergency Management—Public Safety Canada, https://www.publicsafety.gc.ca/cnt/mrgnc-mngmnt/index-en.aspx).

Mitigation

Preventive or mitigation measures take different forms for different types of disasters. In earthquake-prone areas, these preventive measures might include structural changes such as the installation of an earthquake valve to instantly shut off the natural gas supply, seismic retrofits of property, and the securing of items inside a building. The latter may include the mounting of furniture, refrigerators, water heaters, and breakables to the walls and the addition of cabinet latches. In flood-prone areas, houses can be built on poles/stilts. In areas prone to prolonged electricity black-outs, installation of a generator ensures continuation of electrical service. The construction of storm cellars and fallout shelters are further examples of personal mitigation actions.

Disaster mitigation measures are those that eliminate or reduce the impacts and risks of hazards through proactive measures taken before an emergency or disaster occurs.

Preparedness

Preparedness focuses on preparing equipment and procedures for use when a disaster occurs. This equipment and these procedures can be used to reduce vulnerability to disaster, to mitigate the impacts of a disaster or to respond more efficiently in an emergency. The FEMA has set out a basic four-stage vision of preparedness flowing from mitigation to preparedness to response to recovery and back to mitigation in a circular planning process. This circular, overlapping model has been modified by other agencies, taught in emergency class, and discussed in academic papers.

FEMA also operates a Building Science Branch that develops and produces multi-hazard mitigation guidance that focuses on creating disaster-resilient communities to reduce loss of life and property. FEMA advises citizens to prepare their homes with some emergency essentials in the case that the food distribution lines are interrupted. FEMA has subsequently prepared for this contingency by purchasing hundreds of thousands of freeze-dried food emergency meals ready to eat to dispense to the communities where emergency shelter and evacuations are implemented.

Emergency preparedness can be difficult to measure. The Centers for Disease Control and Prevention (CDC) (https://www.cdc.gov) focuses on evaluating the effectiveness of its public health efforts through a variety of measurement and assessment programs.

Local Emergency Planning Committees

Local emergency planning committees (LEPCs) are required by the U.S. Environmental Protection Agency, under the Emergency Planning and Community Right-to-Know Act, to develop an emergency response plan, review the plan at least annually, and provide information about chemicals in the community to local citizens. This emergency preparedness effort focuses on hazards presented by use and storage of extremely hazardous, hazardous, and toxic chemicals. Particular requirements of LEPCs include the following:

- Identification of facilities and transportation routes of extremely hazardous substances
- Description of emergency response procedures, on and off site
- Designation of a community coordinator and facility emergency coordinator(s) to implement the plan
- Outline of emergency notification procedures
- Description of how to determine the probable affected area and population by releases
- Description of local emergency equipment and facilities and the persons responsible for them
- Outline of evacuation plans
- A training program for emergency responders (including schedules)
- Methods and schedules for exercising emergency response plans

According to the Environmental Protection Agency, "Many LEPCs have expanded their activities beyond the requirements of EPCRA, encouraging accident prevention and risk reduction, and addressing homeland security in their communities," and the Agency offers advice on how to evaluate the effectiveness of these committees ("Disaster Management with Case Study," SlideShare, https://www.slideshare.net/.../disaster-management-with-case-study).

Preparedness Measures

Preparedness measures can take many forms, ranging from focusing on individual people, locations, or incidents to broader, government-based "all-hazard" planning. There are a number of preparedness stages between all-hazard and individual planning, generally involving some combination of both mitigation and response planning. Business continuity planning encourages businesses to have a disaster recovery plan. Community- and faith-based organizations' mitigation efforts promote field response teams and interagency planning.

Classroom Response Kit

School-based response teams cover everything from live shooters to gas leaks and nearby bank robberies. Educational institutions plan for cyber-attacks and windstorms. Industry-specific guidance exists for horse farms, boat owners, and more (Industrial Principal Emergency Response and Preparedness, https://www.osha.gov/Publications/osha3122.pdf).

Family preparedness for disaster is fairly unusual. A 2013 survey found that only 19% of American families felt that they were "very prepared" for a disaster. Still, there are many resources available for family disaster planning ("Disaster Management with Case Study," SlideShare Families, https://www.slideshare.net/.../disaster-management-with-case-study). The Department of Homeland Security's* page includes a Family Emergency Plan Checklist; has a whole webpage devoted to readiness for kids, complete with cartoon-style superheroes; and ran a Thunderclap Campaign in 2014. The CDC has a Zombie Apocalypse website (Zombie Preparedness, PHPR, CDC, https://www.cdc.gov/phpr/zombie/index.htm).

Kitchen Fire Extinguisher

Disasters take a variety of forms to include earthquakes, tsunamis, or regular structure fires. That a disaster or emergency is not large scale in terms of population or acreage impacted or duration does not make it any less of a disaster for the people or area impacted, and much can be learned about preparedness from so-called small disasters. The Red Cross states that it responds to nearly 70,000 disasters a year, the most common of which is a single family fire (https://www.redcross.org/ns/apology/disaster_homepage.html?q=red+cross&form=EDNTHT&mkt=en-us&httpsmsn=1&refig=62c7e55a966c4731b4789c17e0ab1aec&sp=1&http%3A%2F%2Fwww.redcross.org%2F=).

Items on Shelves in Basements

Preparedness starts with an individual's everyday life and involves items and training that would be useful in an emergency. What is useful in an emergency is often also useful in everyday life. From personal preparedness, preparedness continues on a continuum through family preparedness, community preparedness, and then business, nonprofit, and governmental preparedness. Some organizations blend these various levels. For example, the International Red Cross and Red Crescent Movement have a webpage on disaster training as well as offer training on basic preparedness such as cardiopulmonary resuscitation and first aid (International Red Cross and Red Crescent Movement, IFRC, official site www.ifrc.org). Other nonprofits such as Team Rubicon bring specific groups of people into disaster preparedness and response operations (Team Rubicon, official site, https://teamrubiconusa.org). FEMA breaks down preparedness into a pyramid, with citizens on the foundational bottom, on top of which rests local government, state government, and federal government in that order ("Tiered Response Pyramid: A System-Wide Approach to Build Response Capability and Surge Capacity," https://www.hsaj.org/article).

Nonperishable Food in Cabinets

The basic theme behind preparedness is to be ready for an emergency, and there are a number of different variations of being ready based on an assessment of what sort

* Ready.gov knows what disasters could affect your area, how to get emergency alerts, and where you would go if you and your family need to evacuate. Check out the related links to learn what to do before, during and after each type of emergency.

of threats exist. Nonetheless, there is basic guidance for preparedness that is common despite an area's specific dangers. FEMA recommends that everyone have a three-day survival kit for their household. Because individual household sizes and specific needs might vary, FEMA's recommendations are not item specific, but the list includes the following:

- Three-day supply of nonperishable food
- Three-day supply of water—one gallon of water per person, per day
- Portable, battery-powered radio or television and extra batteries
- Flashlight and extra batteries
- First aid kit and manual
- Sanitation and hygiene items (e.g., toilet paper, menstrual hygiene products)
- Matches and waterproof container
- Whistle
- Extra clothing
- Kitchen accessories and cooking utensils, including a can opener
- Photocopies of credit and identification cards
- Cash and coins
- Special needs items, such as prescription medications, eyeglasses, contact lens solutions, and hearing aid batteries
- Items for infants, such as formula, diapers, bottles, and pacifiers
- Other items to meet unique family needs

Along similar lines, but not exactly the same, the CDC has its own list for a proper disaster supply kit (https://www.cdc.gov).

- Water—one gallon per person, per day
- Food—nonperishable, easy-to-prepare items
- Flashlight
- Battery-powered or hand crank radio (NOAA Weather Radio, if possible)
- Extra batteries
- First aid kit
- Medications (seven-day supply), other medical supplies, and medical paperwork (e.g., medication list and pertinent medical information)
- Multipurpose tool (e.g., Swiss army knife)
- Sanitation and personal hygiene items
- Copies of personal documents (e.g., proof of address, deed/lease to home, passports, birth certificates, and insurance policies)
- Cell phone with chargers
- Family and emergency contact information
- Extra cash
- Emergency blanket
- Map(s) of the area
- Extra set of car keys and house keys
- Manual can opener

Children are a special population when considering emergency preparedness, and many resources are directly focused on supporting them. The Substance Abuse and Mental Health Services Administration (SAMHSA) (https://www.samhsa.gov) has a list of tips for talking to children during infectious disease outbreaks, to include being a good listener, encouraging children to ask questions, and modeling self-care by setting routines, eating healthy meals, getting enough sleep, and taking deep breaths to handle stress.

FEMA has similar advice, noting that "Disasters can leave children feeling frightened, confused, and insecure," whether a child has experienced it first hand, had it happen to a friend, or simply saw it on television. In the same publication (https://www.fema.gov/disasters), FEMA further notes, "Preparing for disaster helps everyone in the family accept the fact that disasters do happen, and provides an opportunity to identify and collect the resources needed to meet basic needs after disaster. Preparation helps; when people feel prepared, they cope better and so do children."

To help people assess what threats might be in order to augment their emergency supplies or improve their disaster response skills, FEMA has published a booklet called the "Threat and Hazard Identification and Risk Assessment Guide" (THIRA) (https://www.fema.gov/threat-and-hazard-identification-and-risk). This guide, which outlines the THIRA process, emphasizes "whole community involvement," not just governmental agencies, in preparedness efforts.

In this guide, FEMA breaks down hazards into three categories: natural, technological, and human caused, and notes that each hazard should be assessed for both its likelihood and its significance. According to FEMA, "Communities should consider only those threats and hazards that could plausibly occur" and "Communities should consider only those threats and hazards that would have a significant effect on them." To develop threat and hazard context descriptions, communities should take into account the time, place, and conditions in which threats or hazards might occur.

Not all preparedness efforts and discussions involve the government or established nongovernmental organizations like the Red Cross. Emergency preparation discussions are active on the Internet, with many blogs and websites dedicated to discussing various aspects of preparedness. Online sales of items such as survival food, medical supplies, and heirloom seeds allow people to stock basements with cases of food and drinks with 25-year shelf lives, sophisticated medical kits, and seeds that are guaranteed to sprout even after years of storage.

Not all emergency preparation efforts revolve around food, guns, and shelters, although these items help address the needs in the bottom two sections of Maslow's hierarchy of needs ("Maslow's Hierarchy of Needs—Learning Theories," https://www.learning-theories.com/maslows-hierarchy-of-needs.html). The American Preppers Network has an extensive list of items that might be useful in less apparent ways than a first aid kid or help add "fun" to challenging times (American Preppers Network—National Family Survival, americanpreppersnetwork.com). These items include the following:

- Books and magazines
- Arts and crafts

- Children's entertainment
- Crayons and coloring books
- Notebooks and writing supplies
- Nuts, bolts, screws, nails, etc.
- Religious material
- Sporting equipment, card games, and board games

Emergency preparedness goes beyond immediate family members. For many people, pets are an integral part of their families, and emergency preparation advice includes them as well. It is not unknown for pet owners to die while trying to rescue their pets from a fire or from drowning. CDC's Disaster Supply Checklist for Pets includes the following (Disaster Preparedness for Your Pet, Features, CDC, https://www.cdc.gov/features/petsanddisasters/index.html):

- Food and water for at least 3 days for each pet; bowls, and a manual can opener.
- Depending on the pet, you may need a litter box, paper towels, plastic trash bags, grooming items, and/or household bleach.
- Medications and medical records stored in a waterproof container.
- First aid kit with a pet first aid book.
- Sturdy leash, harness, and carrier to transport pet safely. A carrier should be large enough for the animal to stand comfortably, turn around, and lie down. Your pet may have to stay in the carrier for several hours.
- Pet toys and the pet's bed, if you can easily take it, to reduce stress.
- Current photos and descriptions of your pets to help others identify them in case you and your pets become separated and to prove that they are yours.
- Information on feeding schedules, medical conditions, behavior problems, and the name and telephone number of your veterinarian in case you have to board your pets or place them in foster care.

Emergency preparedness also includes more than physical items and skill-specific training. Psychological preparedness is also a type of emergency preparedness, and specific mental health preparedness resources are offered for mental health professionals by organizations such as the Red Cross. These mental health preparedness resources are designed to support both community members affected by a disaster and the disaster workers serving them. CDC has a website (https://www.cdc.gov) devoted to coping with a disaster or traumatic event.

After such an event, the CDC, through the SAMHSA, suggests that people seek psychological help when they exhibit symptoms such as excessive worry, crying frequently, an increase in irritability, anger, and frequent arguing, wanting to be alone most of the time, feeling anxious or fearful, overwhelmed by sadness, confused, having trouble thinking clearly and concentrating, and difficulty making decisions, increased alcohol and/or substance use, increased physical (aches, pains) complaints such as headaches, and trouble with "nerves."

Sometimes, emergency supplies are kept in what is called a bug-out bag. While FEMA does not actually use the term "bug-out bag," calling it instead as some variation of a "go kit," the idea of having emergency items in a quickly accessible place is common to both FEMA and CDC, although online discussions of what items a bug-out bag should include sometimes cover items such as firearms and great knives that are not specifically suggested by FEMA or CDC (https://www.cdc.gov). The theory behind a bug-out bag is that emergency preparations should include the possibility of emergency evacuation. Whether fleeing a burning building or hastily packing a car to escape an impending hurricane, flood, or dangerous chemical release, rapid departure from a home or workplace environment is always a possibility, and FEMA suggests having a family emergency plan for such occasions.

Because family members may not be together when disaster strikes, this plan should include reliable contact information for friends or relatives who live outside of what would be the disaster area for household members to notify they are safe or otherwise communicate with each other. Along with the contact information, FEMA suggests having well-understood local gathering points if a house must be evacuated quickly to avoid the dangers of reentering a burning home. Family and emergency contact information should be printed on cards and put in each family member's backpack or wallet. If family members spend a significant amount of time in a specific location, such as at work or school, FEMA suggests learning the emergency preparation plans for those places.

FEMA has a specific form, in English and in Spanish, to help people put together these emergency plans, although it lacks lines for e-mail contact information (Resources for Other Languages, https://www.fema.gov/resources-other-languages).

Like children, people with disabilities and other special needs have special emergency preparation needs. While "disability" has a specific meaning for specific organizations such as collecting Social Security benefits, for the purposes of emergency preparedness, the Red Cross uses the term in a broader sense to include people with physical, medical, sensor, or cognitive disabilities or the elderly and other special needs populations. Depending on the particular disability, specific emergency preparations might be required.

FEMA's suggestions for people with disabilities includes having copies of prescriptions, charging devices for medical devices such as motorized wheel chairs, and a week's supply of medication readily available or in a "go stay kit." In some instances, lack of competency in English may lead to special preparation requirements and communication efforts for both individuals and responders.

FEMA notes that long-term power outages can cause damage beyond the original disaster that can be mitigated with emergency generators or other power sources to provide an emergency power system. The U.S. Department of Energy states that "homeowners, business owners, and local leaders may have to take an active role in dealing with energy disruptions on their own." This active role may include installing or procuring generators that are either portable or permanently mounted and run on fuels such as propane or natural gas or gasoline.

Concerns about carbon monoxide poisoning, electrocution, flooding, fuel storage, and fire leads even small property owners to consider professional installation and maintenance. Major institutions like hospitals, military bases, and educational institutions often have or are considering extensive backup power systems. Instead of, or in addition to, fuel-based power systems, solar, wind, and other alternative power sources may be used. Standalone batteries, large or small, are also used to provide backup charging for electrical systems and devices ranging from emergency lights to computers to cell phones.

Emergency preparedness does not stop at home or at school. The U.S. Department of Health and Human Services addresses specific emergency preparedness issues that hospitals may have to respond to, including maintaining a safe temperature, providing adequate electricity for life support systems, and even carrying out evacuations under extreme circumstances (www.dmort7.org/downloads/2009SpringNationalNews.pdf). FEMA encourages all businesses to have an emergency response plan, and the Small Business Administration specifically advises small business owners to also focus on emergency preparedness and provides a variety of different worksheets and resources.

FEMA cautions that emergencies happen while people are travelling as well and provides guidance around emergency preparedness for a range of travelers to include commuters, commuter emergency plan, and holiday travelers. In particular, Ready.gov has a number of emergency preparations specifically designed for people with cars. These preparations include having a full gas tank, maintaining adequate windshield wiper fluid, and other basic car maintenance tips. Items specific to an emergency include the following:

- Jumper cables: might want to include flares or reflective triangle
- Flashlights, to include extra batteries (batteries have less power in colder weather)
- First aid kit, to include any necessary medications, as well as baby formula and diapers if caring for small children
- Nonperishable food such as canned food (be alert to liquids freezing in colder weather) and protein-rich foods like nuts and energy bars
- Manual can opener
- At least 1 gallon of water per person a day for at least 3 days (be alert to hazards of frozen water and resultant container rupture)
- Basic toolkit: pliers, wrench, screwdriver
- Pet supplies: food and water
- Radio: battery or hand cranked
- For snowy areas: cat litter or sand for better tire traction; shovel; ice scraper; warm clothes, gloves, hat, sturdy boots, jacket, and an extra change of clothes
- Blankets or sleeping bags
- Charged cell phone and car charger

In addition to emergency supplies and training for various situations, FEMA offers advice on how to mitigate disasters. The Agency gives instructions on how to retrofit a home to minimize hazards from a flood, to include installing a backflow prevention device, anchoring fuel tanks and relocating electrical panels.

Marked Gas Shutoff

Given the explosive danger posed by natural gas leaks, Ready.gov states unequivocally that “It is vital that all household members know how to shut off natural gas” and that property owners must ensure they have any special tools needed for their particular gas hookups. Ready.gov also notes that "It is wise to teach all responsible household members where and how to shut off the electricity," cautioning that individual circuits should be shut off before the main circuit. Ready.gov further states that "It is vital that all household members learn how to shut off the water at the main house valve" and cautions that the possibility that rusty valves might require replacement.

Response

The response phase of an emergency may commence with search and rescue, but in all cases, the focus will quickly turn to fulfilling the basic humanitarian needs of the affected population. This assistance may be provided by national or international agencies and organizations. Effective coordination of disaster assistance is often crucial, particularly when many organizations respond and local emergency management agency capacity has been exceeded by the demand or diminished by the disaster itself. The National Response Framework (NRF) is a U.S. government publication that explains the responsibilities and expectations of government officials at the local, state, federal, and tribal levels. It provides guidance on emergency support functions that may be integrated in whole or parts to aid in the response and recovery process.

On a personal level, the response can take the shape either of a shelter in place or an evacuation.

Evacuation Sign

In a shelter-in-place scenario, a family would be prepared to fend for themselves in their home for many days without any form of outside support. In an evacuation, a family leaves the area by automobile or other mode of transportation, taking with them the maximum amount of supplies they can carry, possibly including a tent for shelter. If mechanical transportation is not available, evacuation on foot would ideally include carrying at least three days of supplies and rain-tight bedding, a tarpaulin, and a bedroll of blankets.

Donations are often sought during this period, especially for large disasters that overwhelm local capacity. Due to efficiencies of scale, money is often the most cost-effective donation if fraud is avoided. Money is also the most flexible, and if goods are sourced locally, then transportation is minimized and the local economy is boosted. Some donors prefer to send gifts in kind; however, these items can end up creating issues, rather than helping. One innovation by Occupy Sandy volunteers is to use a donation registry, where families and businesses impacted by the disaster can make specific requests, which remote donors can purchase directly via a website ("Occupy Sandy: A New Model for Disaster Relief," TIDES, https://startidesnet.wordpress.com/2014/07/09/occupy-sandy-a-new).

Medical considerations will vary greatly based on the type of disaster and secondary effects. Survivors may sustain a multitude of injuries to include lacerations, burns, near drowning, or crush syndrome.

Recovery

The recovery phase starts after the immediate threat to human life has subsided. The immediate goal of the recovery phase is to bring the affected area back to normalcy as quickly as possible. During reconstruction, it is recommended to consider the location or construction material of the property.

The most extreme home confinement scenarios include war, famine, and severe epidemics and may last a year or more. Then recovery will take place inside the home. Planners for these events usually buy bulk foods and appropriate storage and preparation equipment and eat the food as part of normal life. A simple balanced diet can be constructed from vitamin pills, whole-meal wheat, beans, dried milk, corn, and cooking oil. One should add vegetables, fruits, spices, and meats, both prepared and fresh-gardened, when possible.

AS A PROFESSION

Professional emergency managers can focus on government and community preparedness or private business preparedness. Training is provided by local, state, federal, and private organizations and ranges from public information and media relations to high-level incident command and tactical skills.

In the past, the field of emergency management has been populated mostly by people with a military or first-responder background. Currently, the field has become more diverse, with many managers coming from a variety of backgrounds other than the military or first responder fields. Educational opportunities are increasing for those seeking undergraduate and graduate degrees in emergency management or a related field. There are over 180 schools in the United States with emergency-management-related programs, but only one doctoral program specifically in emergency management.

Professional certifications such as Certified Emergency Manager (CEM) and Certified Business Continuity Professional (CBCP) are becoming more common as professional standards are raised throughout the field, particularly in the United States. There are also professional organizations for emergency managers, such as the National Emergency Management Association and the International Association of Emergency Managers.

Principles

In 2007, Dr. Wayne Blanchard of FEMA's Emergency Management Higher Education Project, under the direction of Dr. Cortez Lawrence, superintendent of FEMA's Emergency Management Institute, convened a working group of emergency management practitioners and academics to consider principles of emergency management. This was the first time the principles of the discipline were to be codified. The group agreed on eight principles that will be used to guide the development of a doctrine of emergency management. Below is a summary (mentioned also in Chapter 2):

1. Comprehensive—consider and take into account all hazards, all phases, all stakeholders and all impacts relevant to disasters.
2. Progressive—anticipate future disasters and take preventive and preparatory measures to build disaster-resistant and disaster-resilient communities.

3. Risk-driven—use sound risk management principles (hazard identification, risk analysis, and impact analysis) in assigning priorities and resources.
4. Integrated—ensure unity of effort among all levels of government and all elements of a community.
5. Collaborative—create and sustain broad and sincere relationships among individuals and organizations to encourage trust, advocate a team atmosphere, build consensus, and facilitate communication.
6. Coordinated—synchronize the activities of all relevant stakeholders to achieve a common purpose.
7. Flexible—use creative and innovative approaches in solving disaster challenges.
8. Professional—value a science and knowledge-based approach; based on education, training, experience, ethical practice, public stewardship and continuous improvement (Principles of Emergency Management Supplement, FEMA.gov, https://www.fema.gov/media-library-data/20130726-1822-25045-7625/principles_of_emergency_management.pdf).

Tools

In recent years, the continuity feature of emergency management has resulted in a new concept, Emergency Management Information Systems (EMIS). For continuity and interoperability between emergency management stakeholders, EMIS supports an infrastructure that integrates emergency plans at all levels of government and nongovernment involvement for all four phases of emergencies. In the healthcare field, hospitals utilize the Hospital Incident Command System (emsa.ca.gov, http://www.emsa.ca.gov/media/default/HICS/HICS_Guidebook_2014_10.pdf; HICS is an incident), which provides structure and organization in a clearly defined chain of command.

Disaster Response Technologies

A Smart Emergency Response System (SERS) prototype was built in the Smart America Challenge 2013–2014, a U.S. government initiative. SERS has been created by a team of nine organizations led by MathWorks (Smart America, smartamerica.org/teams/smart-emergency-response-system-sers).

The Smart America initiative challenges the participants to build cyber-physical systems as a glimpse of the future to save lives, create jobs, foster businesses, and improve the economy. SERS primarily saves lives. The system provides the survivors and the emergency personnel with information to locate and assist each other during a disaster. SERS allows submission of help requests to a MATLAB-based mission center connecting first responders, apps, search-and-rescue dogs, a 6-feet-tall humanoid, robots, drones, and autonomous aircraft and ground vehicles.

The command and control center optimizes the available resources to serve every incoming request and generates an action plan for the mission. The Wi-Fi network is created on the fly by drones equipped with antennas. In addition, the autonomous rotorcrafts, planes, and ground vehicles are simulated with Simulink and visualized in a 3D environment (Google Earth) to unlock the ability to observe the operations on a mass scale.

Within Other Professions

Practitioners in emergency management come from an increasing variety of backgrounds. Professionals from memory institutions (e.g., museums, historical societies, etc.) are dedicated to preserving cultural heritage—objects and records. This has been an increasingly major component within this field as a result of the heightened awareness following the September 11 attacks in 2001, the hurricanes in 2005, and the collapse of the Cologne Archives.

To increase the potential successful recovery of valuable records, a well-established and thoroughly tested plan must be developed. This plan should emphasize simplicity in order to aid in response and recovery: employees should perform similar tasks in the response and recovery phase that they perform under normal conditions. It should also include mitigation strategies such as the installation of sprinklers within the institution. Professional associations hold regular workshops to keep individuals up to date with tools and resources in order to minimize risk and maximize recovery.

Absolute requirements for participation are as follows:

- Have I chosen to participate?
- Have I taken Incident Command System (ICS) training?
- Have I taken other required background courses?
- Have I made arrangements with my company to deploy?
- Have I made arrangements with my family?

Incident participation:

- Have I been invited to participate?
- Are my skill sets a match for the mission?
- Can I access just-in-time training to refresh skills or acquire needed new skills?
- Is this a self-support mission?
- Do I have supplies needed for three to five days of self-support?

FEMA's Emergency Management Institute

The Emergency Management Institute's (EMI's) main campus is in Emmitsburg, Maryland.

The EMI serves as the national focal point for the development and delivery of emergency management training to enhance the capabilities of state, territorial, local, and tribal government officials; volunteer organizations; FEMA's disaster workforce; other federal agencies; and the public and private sectors to minimize the impact of disasters and emergencies on the American public. EMI curricula are structured to meet the needs of this diverse audience with an emphasis on separate organizations working together in all-hazards emergencies to save lives and protect property. Particular emphasis is placed on governing doctrine such as the NRF, National Incident Management System, and the National Preparedness Guidelines. EMI is fully accredited by the International Association for Continuing Education and Training and the American Council on Education.

The Independent Study (IS) program at EMI consists of free courses offered to U.S. citizens in Comprehensive Emergency Management techniques. Course IS-1 is entitled "Emergency Manager: An Orientation to the Position" and provides background information on FEMA and the role of emergency managers in agency and volunteer organization coordination. The EMI IS Program, a web-based distance learning program open to the public, delivered extensive online training with approximately 200 courses and trained more than 2.8 million individuals. IS courses are online and at no cost to anyone who has the proper login information and SSID student number (e.g., 0000123456).

2 Introduction to Emergency Management: Step by Step Process

PLANNING PROCESS

Definitions

- Hazard refers to the potential for extreme events to affect people, property and the natural environment in a given location.
- Two descriptions of emergency are as follows:
  - Minor events that cause a few casualties and a limited amount of property damage.
  - An imminent event that will likely strike soon.
- Disaster is reserved for events that produce more losses than a community can handle.

Natural Disasters

- Large scale natural disasters are common (earthquakes, floods, hurricanes, volcanic eruptions, and wild land fires).
- More people are being affected by natural disasters and losses are becoming progressively greater.
- People are choosing hazardous building designs and inadequate structural materials that fail under extreme stress.

Technological Disasters

- People are more affected now by technological disasters because there are more people living close to them.
- Risks are rising from increased quantity and variety of hazardous materials.
- Use of nuclear power plants and liquefied natural gas facilities.

Terrorist Disasters

- Terrorists use some of the same materials involved in technological disasters and are intended to cause many casualties and major damage.

- Emergency managers respond to terrorist attacks using the same basic approach used in other disasters.
- Emergency managers must
  - Rapidly detect and assess the situation.
  - Mobilize relevant organizations and facilities.
  - Take action to limit casualties and damage.
  - Coordinate the organizations responding to the incident.

Emergency Manager's Role

- The emergency manager's role is to prevent or reduce losses that occur due to hazards, disasters, and emergencies.
- Losses are measured in a variety of ways, including the following:
  - Number of deaths and injuries
  - Property damage
- Increased importance of emergency management:
  - Public awareness of hazards and disasters has increased along with the cost of disasters.
  - Businesses can be disrupted and even go bankrupt.
  - Rapid population growth in hazardous geographical areas has created increased disaster exposure.
  - Emergency manager's training has led to the development of emergency management as a profession.

Local Management

- Emergency management is a local job to influence events with local consequences.
- The Federal Emergency Management Agency's (FEMA's) practice of attempting to manage events locally whenever possible places a major burden on state and local government agencies.
- The current National Response Plan states that local jurisdictions must be able to operate without external help for 72 hours after hazard impact (https://www.dhs.gov/xlibrary/assets/NRP_Brochure.pdf).

Planning Process

There are four steps in performing a hazard/vulnerability analysis:

- Identify hazards.
- Estimate the probability.
- Project the consequences.
- Communities need to develop hazard management strategies from four basic elements: hazard mitigation, emergency preparedness, emergency response, and disaster recovery.

Hazard Mitigation

- Hazard mitigation addresses the causes of a disaster, reducing the likelihood it will occur or limiting its impact.
- Focus is to stop disasters before they happen.
- Changing either the natural event or human behavior, or both, reduces the impact of a natural event such as a flood, hurricane, or earthquake.
- The choice of whether to reduce technological hazards by controlling the hazard agent or by controlling the human use depends on political and economic decisions.

Preparedness

- Preparedness consists of plans, procedures, and resources that must be developed in advance.
- Designed to support a timely and effective emergency response to an imminent impact and guide the disaster recovery process.
- Disaster program needs to answer four questions:
  - Which agencies will participate?
  - What response and recovery actions are feasible?
  - How will the response and recovery organizations function and what resources are needed?
  - How will disaster preparedness be established and maintained?

Response

- Emergency response begins when the event occurs or when hazard-monitoring systems alert authorities of an imminent disaster.
- Emergency response goals are as follows:
  - Protecting the population.
  - Limiting damage from the primary impact.
  - Minimizing damage from secondary impacts.
- Secondary impacts are "disasters caused by the disaster" and include hazardous materials releases initiated by earthquakes.
- Local emergency responders dominate the response period, which is characterized by uncertainty and urgency.

Recovery

- Recovery begins as the disaster is ending and continues until the community is back to normal.
- The immediate goal is to restore the community's infrastructure.
- The ultimate goal is to return the community's quality of life to the same level as it was before the disaster.
- Most resources during recovery come from outside the community—the majority of resources in a major disaster come from the federal government.

Strategy Mix

- Historically, more emergency management resources are used for response and recovery than for mitigation and preparedness.
- Recent events, such as Hurricane Katrina, have called attention to the flaw in this policy.
- Community's hazard management strategy should combine all four activities.

Summary

- This section demonstrated how you must prepare for threats through the use of mitigation, preparedness, response, and recovery plans. The use of technological systems, risk communication, and sanctions and incentives enables the effective implementation of emergency plans.

Key Terms

- Disaster
- Emergency
- Hazard
- Mitigation
- Natural disaster
- Recovery
- Response
- Secondary impacts
- Terrorist disasters
- Technological disasters

THE FOUR PILLARS OF EMERGENCY MANAGEMENT

Note: No community is exempt from emergencies.

- As much as we dislike thinking about the potential for natural, technological, human-related disasters, or disasters involving hazardous material, these events do occur.
- The best way to minimize the effects of such events is to discuss the potential vulnerabilities that exist in an organization, while instituting strategies that include the components of mitigation, preparedness, response, and recovery (Figure 2.1).
- On August 3, 1993, President Clinton signed Executive Order 12856, "Federal Compliance with the 'Right-to-Know' Laws and Pollution Prevention Requirements" (National Archives, https://www.archives.gov/.../executive-orders/pdf/12856.pdf).
- This Executive Order requires federal agencies, including the Department of Defense, to fully comply with all provisions of Emergency Planning and Community Right-to-Know Act (EPCRA) and the Pollution Prevention Act

FIGURE 2.1 Emergency management planning tree.

(https://dps.mn.gov/divisions/hsem/epcra) with one notable exception: the reporting requirements under Section 313.

- Under the EPCRA, State Emergency Response Commission and Local Emergency Planning Committee are charged with four primary responsibilities:
  - Write emergency plans to protect the public from chemical accidents.
  - Establish procedures to warn and, if necessary, evacuate the public in case of an emergency.
  - Provide citizens and local governments with information about hazardous chemicals and accidental releases of chemicals in their communities.
  - Assist in the preparation of public reports on annual release of toxic chemicals into the air, water, and soil.
- *The plan must be reviewed annually*, more often is preferred, to include a table-top exercise, a small local exercise, and a full scale exercise to see if the plan is functional.

The four pillars of emergency management are the phases of planning and action undertaken to ensure a comprehensive approach to emergency management, while maximizing the security of the infrastructure, continuity of operations, and the safety of staff, public.

The four pillars are as follows (Figure 2.2).

Mitigation

- Mitigation is the most cost-efficient method for reducing the impact of hazards.
- A precursor activity to mitigation is the identification of risks.
- Physical risk assessment refers to the process of identifying and evaluating hazards.
- The higher the risk, the more urgent the need is to target hazard specific vulnerabilities through mitigation efforts.

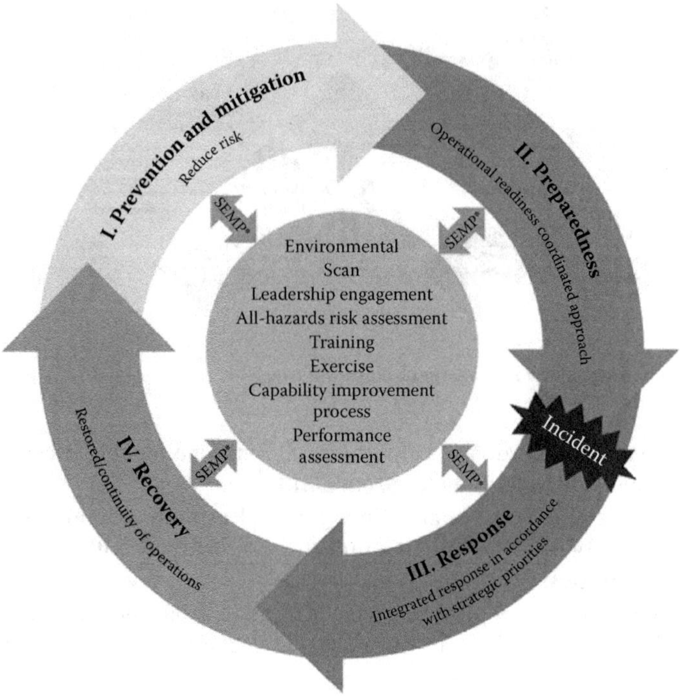

FIGURE 2.2 The four pillars of emergency management.

- One example of mitigation is the 96-Hour Continuity Plan, which includes mitigation strategies and plans that have been developed to ensure continuity of operations in areas such as utilities, communications, food, water, medical support, staffing, and medical supplies when the installation is unable to support due to an external disaster scenario (Emergency Management and Business Continuity, dp.ccalac.org/Policies/BusinessContinuityPlan/Documents/EOP-BCP).

Preparedness

- Preparedness is a continuous cycle of planning, organizing, training, equipping, exercising, evaluation, and improvement activities that ensure effective coordination and the enhancement of capabilities to prevent, protect against, respond to, recover from, and mitigate against disaster events that have been identified within the Hazard Vulnerability Analysis (HVA) (https://training.fema.gov/emiweb/downloads/untservicelearning.pdf).
- In the preparedness phase, the Emergency Management Department develops plans of action to manage and counter risks and takes action to build the necessary capabilities needed to implement such plans.

Response

- The response phase includes the mobilization of the identified emergency staff, including first responders, to an internal or external event that could have an impact on patient care operations.
- Response procedures are predetermined by the local government and are detailed in disaster plans during the preparedness phase.
- Response to an internal or external event on the local government is directed through the Incident Command System (ICS).
- Response plans remain flexible in nature due to the varying members of staff available on the installation at any given time.
- Response procedures and plans are constantly evaluated and changed based on improvements identified during After Action Reviews, which are held after disaster events.
- Response is also evaluated regularly by the local government through drills, exercises, tracers, and live events.

Recovery

- The aim of the recovery phase is to restore the affected area to its previous state.
- It differs from the response phase in its focus: recovery efforts are concerned with issues and decisions that must be made after immediate needs are addressed.
- Recovery efforts are primarily concerned with actions that involve rebuilding destroyed property, reemployment, the repair of other essential infrastructure, as well as the reopening of essential services.
- Recovery operations are an extremely important phase in the emergency management continuum and yet one that is often overlooked.
- The ICS team is responsible for the implementation of the recovery phase.

Hazard Vulnerability Analysis

- The basis of the "all-hazards" approach starts with the installation of the HVA.
- The HVA identifies disasters and other events from a technological, natural, manmade, and hazardous materials perspective that are most prevalent for the region.
- These events are ranked in order of severity and greatest impact to the local government and its operations.
- A risk factor is obtained for each identified hazard by ranking probability, human impact, property impact, business impact, and overall preparedness from an internal and external response entity.

Hazard Vulnerability Analysis Assessment

- The HVA is reviewed annually, or as required by installation leadership and the Emergency Management Committee.
- The HVA, including the top five ranked disasters, is shared with community government and emergency response agencies including the Office of Emergency Management, Public Health, emergency medical services (EMS), police, fire, and the medical examiner's (ME) office.
- The top five events identified by installation were patients (surge), pandemic/epidemic, snowfall, water failure, hurricanes, tornadoes, snow storms, floods and internal flood.

THE PRINCIPLES OF EMERGENCY MANAGEMENT

1. Comprehensive—consider and take into account all hazards, all phases, all stakeholders and all impacts relevant to disasters.
2. Progressive—anticipate future disasters and take preventive and preparatory measures to build disaster-resistant and disaster-resilient communities.
3. Risk-driven—use sound risk management principles (hazard identification, risk analysis, and impact analysis) in assigning priorities and resources.
4. Integrated—ensure unity of effort among all levels of government and all elements of a community.
5. Collaborative—create and sustain broad and sincere relationships among individuals and organizations to encourage trust, advocate a team atmosphere, build consensus, and facilitate communication.
6. Coordinated—synchronize the activities of all relevant stakeholders to achieve a common purpose.
7. Flexible—use creative and innovative approaches in solving disaster challenges.
8. Professional—value a science and knowledge-based approach, based on education, training, experience, ethical practice, public stewardship, and continuous improvement (Principles of Emergency Management Supplement, FEMA.gov, https://www.fema.gov/media-library-data/20130726-1822-25045-7625/principles_of_emergency_management.pdf).

Tools Used in Emergency Management

- In recent years, the continuity feature of emergency management has resulted in a new concept, *Emergency Management Information Systems* (EMIS).
- For continuity and interoperability between emergency management stakeholders, EMIS supports an infrastructure that integrates emergency plans at all levels of government and nongovernment involvement for all four phases of emergencies.
- In the healthcare field, hospitals utilize the hospital ICS, which provides structure and organization in a clearly defined chain of command.

- Professional certifications such as Certified Emergency Manager (CEM) and Certified Business Continuity Professional (CBCP) are becoming more common as professional standards are raised throughout the field, particularly in the United States.
- There are also professional organizations for emergency managers, such as the National Emergency Management Association and the International Association of Emergency.

Emergency Management Committee

- The Emergency Management Committee is a multidisciplinary installation team that is charged with developing processes, policies, and procedures; conducting staff education; and securing necessary resources to ensure a prompt, coordinated, and effective response by local government to all disasters affecting the environment, continuity of operations, life safety, and property of the installation.
- The Committee is responsible for all facets of the emergency operations plan (EOP), including the incorporation of exercises and After Action Reports intended to examine our response while identifying opportunities for improvement.

As a basis for its foundation, the Emergency Management Committee centers its attention on six critical planning areas:

1. Communications
2. Resources and assets
3. Safety and security
4. Staff responsibility
5. Utilities
6. Support activities

These are just six areas to be a foundation of response to a disaster or potential disaster and key elements to an all-hazards approach to emergency management.

Members of the Emergency Management Committee represent multiple disciplines across the installation, including the following:

- Post commander
- Installation commander
- Emergency manager
- Emergency management team
- Emergency operation center (EOC) personnel
- Provost marshal
- Fire department (on and off base)
- Emergency medical service (on and off base)
- Department of Public Works (on and off base)
- Information management and technology

These are just some areas to be considered during the development, planning, writing, and maintain of the EOP.

Comprehensive Emergency Management Plan

- The Comprehensive Emergency Management Plan (CEMP) is a procedural document that contains disaster response plans and policies that assist the organization in applying the all-hazards approach in emergency management.
- The CEMP manual, which is available to installation staff in both paper and electronic formats, contains detailed plans addressing the steps to be taken during specific emergencies.
- The plan defines the roles and responsibilities of different departments and personnel.
- It provides job responsibilities for the different members of the Incident Command Team, listed in job action sheets.
- It includes phone numbers and emergency contacts at the local government and throughout the community.
- The CEMP guides response while providing the Incident Command Team with the information they need to make effective decisions during emergency situations.

Note: The plan must be reviewed annually; this is not an option.

Incident Command System

- The ICS was created in response to a series of destructive wildfires in southern California in 1970 called Fire Scope (for Fire Scope history, see http://www.firescope.org/history.htm).
- National, state, and local emergency responders required a tool to effectively provide command and control functions while coordinating the efforts of many individual agencies and departments working together to stabilize a disaster, while protecting life, property and the environment.

The ICS is used by emergency responders across the nation as well as by our local governments and agencies. The ICS is successful because it uses a common organizational structure with standardized management principles.

- The structure being used in the hospital is similar to the structure being used in the community by police, fire fighters, homeland security, and other rescue personnel.
- With ICS in effect, the hospital and other organizations can communicate effectively and efficiently to solve problems and get the emergency under control.
- In fact, the U.S. Department of Homeland Security requires all organizations that receive federal funds become National Incident Management System (NIMS) compliant, which includes the incorporation of ICS (https://www.fema.gov/national-incident-management-system).

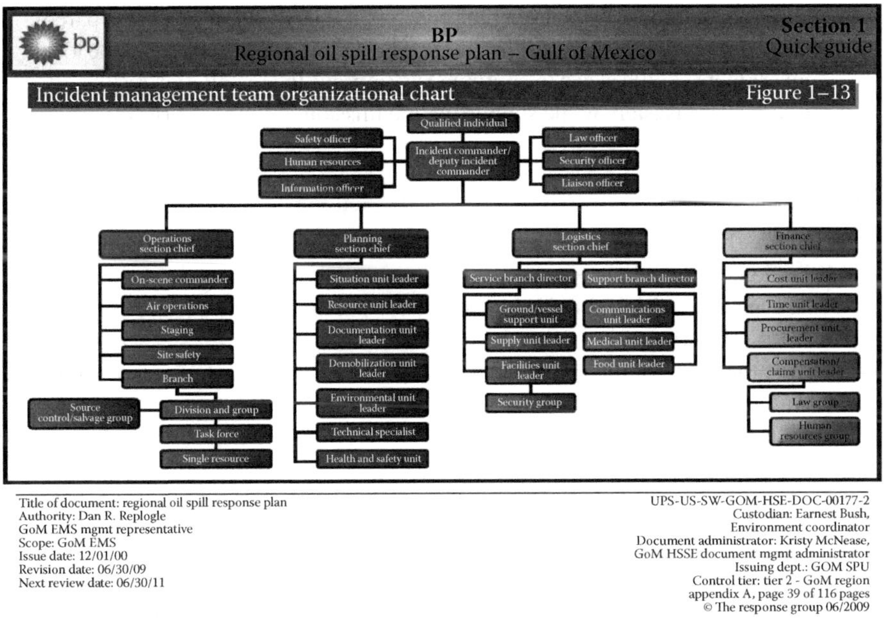

FIGURE 2.3 Example ICS from the BP oil spill.

The ICS has evolved over the years and has taken on a specific meaning for installation systems in the form of the installation ICS (IICS). While IICS provides a prescriptive model for implementation, the installation can tailor the conventional methodology of the ICS in most cases to fit their specific needs (Figure 2.3).

- IICS is employed at the installation to respond to emergency situations. When a disaster is declared, IICS is "activated" and an entire system of planning and operations is put into play.
- Chains of command and job descriptions in the form of specific job action sheets developed by the Emergency Management Department provide everyone on the IICS team with the information they need and the procedures to follow to address the emergency situation at hand.

The Department of Emergency Management at the installation should specialize in this type of "what if" thinking.

- The mission is to give the local government personnel the support system they need to continue to provide continuity of operations across all post functions, even in disaster situations.
- In order to offer this critical response, the department maintains an extensive CEMP that addresses an all-hazards approach to disaster preparedness and an emergency management team that continually reviews and improves the program.

- Through a continuous planning approach that includes mitigation, preparedness, response, and recovery, the emergency management program at the local government should strive to continuously improve the installation response to all types of disasters while supporting the mission in times of crisis.

Review

Q: What are the four pillars of emergency management?
A: Mitigation, preparedness, response, recovery
Q: What type of thinking should the emergency planning team have?
A: "What if" thinking
Q: What is the name of the process that the all-hazards approach planning done by?
A: Hazard vulnerability analysis
Q: How often must the EOP be reviewed?
A: Annually

EMERGENCY MANAGEMENT STAKEHOLDERS

Stakeholders

Community stakeholders can be divided into three categories:

- Social groups
- Economic groups
- Political groups

Much of your skill in dealing with stakeholders will depend on power, how it plays in your relationships with stakeholders and how it influences policies.

Social Groups

The basic social group unit is the household. Households

- Try to prevent accidents
- Prepare for natural disasters
- Evacuate
- Suffer economic losses

Community emergency response teams (CERTs) help homeowners organize as groups to develop emergency management policy.

Economic Groups

- Small businesses are the most at risk, but are more likely to respond to appeals for assistance.
- Businesses tend to focus on response and recovery.
- Disasters cause business interruption—a loss of revenue due to a disruption.
- A critical business stakeholder is public utilities.

Government Stakeholders

Governmental stakeholders include the town or city, the county, the state, and the federal level.

The most important stakeholders are the state emergency management agencies (SEMAs). They

- Set policies
- Provide the legal framework
- Link local governments with FEMA

At the national level, with the establishment of Health and Safety Authority, a restructuring of emergency management began. FEMA's role remains to be determined.

Involving Stakeholders

The local emergency manager must involve all relevant stakeholders in the process.

There are four simple ways to get the community involved in hazard prevention.

- Talk up your work to friends and neighbors. Creating buzz about emergency management is an inexpensive and valuable way to get community support.
- Set up a hazard hotline. Advertise the hotline.
- Speak at schools, neighborhood, and community organizations. Discuss potential threats and emergency management plans.
- Form citizen committees to help advise on the emergency management plans and to get volunteers for carrying out the plans.

Power

Stakeholders have different levels of power.

Organizational theorists have described six bases of power:

- Reward
- Coercive
- Legitimate
- Expert
- Referent
- Information power

Power operates in the upward and downward directions. Power operates along a chain between households and the federal government.

Policy Process

- The basic steps of the policy process are five stages through which policies move.
- The emergency manager's first task is to put hazards on the political agenda.

- Three types of political agendas are as follows:
  - The systemic agenda is hot topics that concern voters.
  - The governmental agenda is the issues the government is working on.
  - The institutional agenda is the issues that institutions are working on.
- A natural or technological disaster is a focusing event that draws public attention to the need for local disaster planning and hazard mitigation.

Agenda Setting

- Window of opportunity—The time immediately following a focusing event during which local emergency managers are most likely to be able to influence policy.
- Because of the short amount of time to effect policy change, individual actors must work aggressively to set issues on the agenda and to keep them there.
- Three qualities are needed to be successful:
  - Technical expertise in hazards, acquired either through education or experience
  - Political expertise necessary for any successful policy change effort
  - Personal commitment because it can take years to overcome opposition to new policies

Policy Formulation

- Managers should have a set of proposals before they attempt to shape the agenda.
- A goal is to minimize court challenges and unintended consequences of the policy or law.
- When developing any public policy, stakeholders must be included. This is especially important for hazard policies, because these policies often require
  - Certain present investment
  - Certain opportunity cost
  - Uncertain future benefit

Policy Adoption

- Policy adoption involves getting stakeholders to urge elected officials to pass the policy.
- Adoption is not the end of the story. All policies must be implemented to be effective.
- Three factors that affect policy implementation.
  - How easy is it to solve the problem?
  - Is there a clear link between the solution and the problem?
  - What level of technology and amount of resources is available to solve the problem?
- A policy should be evaluated periodically and either improved or terminated.

Summary

- This section examined the essential elements of an emergency management policy. It also demonstrated how you can involve stakeholders in emergency management planning and processes. It is important to recognize your own power in influencing stakeholders and policies.

Key Terms

- Business interruption
- CERTs
- Economic groups
- Expert power
- Focusing event
- Governmental groups
- Information power
- Legitimate power
- Referent power
- Reward and coercive power
- Stakeholder
- Social groups
- Window of opportunity

BUILDING AN EFFECTIVE EMERGENCY MANAGEMENT ORGANIZATION

Local Emergency Management Agency

- In practice, a local emergency management agency (LEMA) might be known by as the Office of Civil Defense, Emergency Management, Emergency Services, or Homeland Security.
- The LEMA might be separate or part of another department or an individual attached to the chief administrative officer's (CAO's) office.
- In many communities, one person staffs the LEMA.

Emergency Manager

- As the local emergency manager, you will report to the CAO during emergencies.
- Your status depends on the following:
  - Size of community
  - Financial resources of the community
  - Community's vulnerability to hazards
- You will work mostly with the police, fire, EMS, and public works departments.
- Usually, the CAO is not an expert in public safety, emergency medicine, or emergency management.

Local Emergency Management Agencies

- Some LEMAs have professional staff and many have volunteers with valuable skills.
- Performance reviews are valuable in improving performance effectiveness.
- FEMA has advised emergency managers to set annual goals for major areas.
- A LEMA's budget categorizes anticipated expenses, details each category of funding, and covers the 12 month fiscal year.
- A budget's contingency fund addresses the funds needed in case of an emergency.

Funding Sources

- Your most obvious source of funding is the CAO.
- FEMA has a range of programs that provide financial assistance.
- Continued financial support is based on meeting performance and financial requirements.
- For some forms of FEMA funding, LEMAs submit applications through their SEMAs.
- Industrial facilities, such as nuclear power plants, can help defray the costs of preparing their facilities for an emergency.

Effective Organizations

- An effective LEMA focuses on
    - Organizational outcomes
    - Individual outcomes
    - Planning process
- Frequent, recent, and severe impacts can lead to a disaster subculture in which residents adopt routines to prepare for disasters.
- Community support from senior officials, news media, and public affects the resources allocated to the LEMA.
- Effectiveness of LEMAs can be partly attributed to availability of resources.

LEMA Effectiveness

- Emergency management network effectiveness is greater in communities with recent disaster experience. The more effective networks
    - Have members with more experience
    - Have a wider range of local contacts
    - Have written plans and were familiar with them
    - Have personal experience in managing routine natural hazards such as floods
    - Are more familiar with the policies and procedures of emergency-relevant state and federal agencies
    - Have representation by elected officials and by citizens' groups

Planning Process

The planning process consists of five functions.

1. Plan activities: Cities with a better planning process adopt more preparedness practices.
2. Provide a positive work climate: Planning effectiveness is highest in positive organizational climates.
3. Analyze the situation: Examine the five factors.
4. Acquire resources: Obtain staff, equipment, and information from a variety of sources.
5. Choose a strategy: You can use multiple strategies, and the extent to which you use each one will depend on the size of the community, available funding, and your own personal characteristics.

Outcomes

- For a LEMA to be effective, you need dedicated individuals. People are committed to organizations that provide benefits:
  - Personal (salary, benefits, or feeling of doing something positive for the community)
  - Social (having friends at work)
  - Purposeful (work gives a purpose, an identity)
- Emergency management organizations are often judged based on
  - Effectiveness of plans
  - Timeliness
  - Cost of plans
  - Hazard vulnerability analyses
  - Public information briefings
  - Education efforts, such as brochures and websites

Emergency Operations Plan

The development of an EOP is a multistage process that encompasses eight steps.

- Step 1: Establish a preliminary planning schedule.
  - Identify the principal tasks and the expected amount of time required to perform them.
- Step 2: Publish a planning directive.
  - Coordinate the efforts of other agencies with CAO's delegated authority.
- Step 3: Organize the LEMC.
  - You build a committee by asking others to serve.

  LEMCs are more effective when members are assigned to specific tasks rather than having everyone contribute to all tasks—use subcommittees.
- Step 4: Assess disaster demands and capabilities.
  - LEMC members must identify the tasks that need to be performed in a community-wide emergency.

- Step 5: Write plans.
  - Every LEMA needs an EOP, a recovery operations plan (ROP), and a hazard mitigation plan.
- Step 6: Evaluate and revise draft plans.
  - Make sure that all drafts are reviewed by committees within the LEMC to identify potential problems.
- Step 7: Obtain community review.
  - After the draft plans have been reviewed within the LEMC, release the plans for review throughout the community.
- Step 8: Publish plans in final form.
  - Give all input from the community to the appropriate committees.
  - The committees must address any problems in the final versions of the plans.
  - Forward copies of the final plans and accompanying documents to all government agencies and other participating organizations.

Summary

- LEMAs should be organized and managed well, as should the tasks of developing emergency organizations, and creating emergency operating plans. All of these tasks are outcomes based and all of them will rely on your ability to organize resources, find funding, and communicate well.

Key Terms

- Contingency fund
- Disaster subculture
- LEMA

RISK PERCEPTION AND COMMUNICATION

Risk and Warning

- Risk is a possibility that people or property could be hurt.
  - This risk must be effectively communicated to the people who are likely to be affected.
- A warning is a risk communication about an imminent event that is intended to produce an appropriate disaster response.
- People must receive, heed, and comprehend information about risks.

Warning Processing

- Decisions about how to respond to a hazard begin with risk identification.
  - Important sources are warning messages from authorities, the media, and peers.

- Risk assessment involves evaluating what the personal consequences will be if the disaster occurs.
  - Assessment includes the perceived probability, magnitude, and immediacy of the disaster impact.
- In a protective action search, people are likely to recall actions they have taken on previous occasions.
  - When disaster impact is imminent, households must rely mostly on their own resources to achieve protection.
- Protective action is unlikely to be followed unless it is considered to be effective.
  - The result of protective action assessment is an adaptive plan.
- Protective action implementation occurs when those at risk know they have to take action.
  - People sometimes postpone taking action, even when faced with danger.
- People frequently seek additional information because the consequences of a decision error are very serious.
  - Any confusing messages from officials or any doubts expressed will cause people to seek more information.
- Uncertainty about risk identification and assessment can stimulate questions directed to officials and the news media.
  - People are often forced to rely on the media and their peers even when they would prefer to contact authorities.
- Communication action implementation can have one of three outcomes.
  - First, people can confirm the threat and proceed to take protective action.
  - Second, if an information source is unavailable, people can try to find different sources.
  - Third, if new information contradicts previous information, then people can try to resolve the conflict.

Continuing Hazard Phase

- The continuing hazard phase is marked by a low probability that a catastrophic incident will threaten public safety, property, and the environment.
- During this phase, you should engage in hazard mitigation, emergency preparedness, and recovery preparedness actions.
- There are five basic risk communication functions to address in the continuing hazard phase. These are
  - Strategic analysis
  - Operational analysis
  - Resource mobilization
  - Program development
  - Program implementation

Strategic Analysis

- You should know basic information about your community's ethnic composition, communication channels, perception of authorities, levels of education, and income distribution.
- The hazards that produce the greatest community conflict are those that have a potential for inflicting significant harm on bystanders.
- To be successful, your risk communication program must foster people's sense of personal responsibility for self-protection.
    - Recognize that even the most effectively implemented risk communication program may not lead a large percentage of people to take immediate protective action.

Operational Analysis

- There are five tasks to complete when performing an operational analysis.
    - Task 1: Identify and assess feasible hazard adjustments for the community and its households/businesses.
    - Task 2: Identify ways to provide incentives, sanctions, and technological innovations.
    - Task 3: Identify the available risk communication sources in the community.
    - Task 4: Identify the available risk communication channels in the community.
    - Task 5: Identify specific audience segments.

Resource Mobilization

- As an emergency manager, there are five tasks you must perform to mobilize resources for risk communication.
    - Task 1: Obtain the support of senior appointed and elected officials.
    - Task 2: Enlist the participation of other government agencies.
    - Task 3: Enlist the participation of nongovernmental and private sector organizations.
    - Task 4: Work with the mass media.
    - Task 5: Work with neighborhood associations and civic organizations.

Program Development

- There are five tasks in the program development step.
    - Task 1: Staff and train a crisis communication team.
    - Task 2: Establish procedures for maintaining an effective communication flow during an escalating crisis or emergency response.
    - Task 3: Develop a comprehensive risk communication program.
    - Task 4: Plan to make effective use of informal communication networks.
    - Task 5: Establish procedures for obtaining feedback from the news media and the public.

Program Implementation

- There are five tasks you must complete to implement the risk communication program.
  - Task 1: Build source credibility by increasing perceptions of expertise and trustworthiness.
  - Task 2: Use a variety of channels to disseminate hazard information.
  - Task 3: Describe community or facility hazard adjustments being planned or implemented.
  - Task 4: Describe feasible household hazard adjustments.
  - Task 5: Evaluate program effectiveness.

Escalating Crisis Communication

- An escalating crisis is a situation in which there is a significantly increased probability of an incident occurring that will threaten the public's health, safety, or property.
- Be sure to classify the situation. The categories in the emergency classification system correspond to meaningful differences in the levels of response by local authorities.
- With this system, decisions are made based on rational scientific considerations rather than emotion or other considerations.

Risk Communication

- One of the most important incident management actions is risk communication, which consists of six tasks.
  - Task 1: Activate the crisis communication team promptly.
  - Task 2: Determine the appropriate time to release sensitive information.
  - Task 3: Select the communication channels that are appropriate to the situation.
  - Task 4: Maintain source credibility with the news media and the public.
  - Task 5: Provide timely and accurate information about the hazard to the news media and the public.
  - Task 6: Evaluate performance through post-incident critiques.

Summary

- You learned how critical communication is when it comes to influencing people's perceptions of danger. Apply the communication skills discussed in this lesson, and you just might save lives.

Key Terms

- Adaptive plan
- Escalating crisis
- Risk
- Risk assessment
- Warning

PRINCIPAL HAZARDS IN THE UNITED STATES

Environmental Hazards

- Environmental hazards are commonly classified as natural or technological:
  - Natural hazards are extreme events that originate in nature. Natural hazards are commonly categorized as meteorological, hydrological, or geophysical.
  - Technological hazards originate in human-controlled processes but are released into the air and water. The most important technological hazards include explosives, flammable materials, toxic chemicals, radiological materials, and biological hazards.

Meteorological Hazards

- Severe storms have wind speeds exceeding 58 miles per hour, produce tornadoes, release hail, spawn lightning strikes, and cause flash floods.
- Severe winter storms can immobilize travel, isolate residents of remote areas, deposit enormous amounts of snow that collapse long-span roofs, and bring down telephone and electric power lines.
- Heat associated with extreme summer weather can be a silent killer. Heat-related illnesses include heat cramp, heat syncope, heat exhaustion, and heat stroke.
- An increased number of tornadoes have been reported during recent years, suggesting that some long-term changes in climate are also involved.
- For protection from tornadoes, building occupants should go to an interior room on the lowest floor. Mobile home residents should go to a community shelter.
- Hurricanes produce four specific threats:
  - High winds
  - Tornadoes
  - Inland flooding
  - Storm surge

Wildfires

- Wild land fires burn areas with nothing but natural vegetation for fuel.
- Interface fires burn into areas containing a mixture of natural vegetation and built structures.

- Firestorms are distinguished from other wildfires because they burn so intensely that they warrant a special category.
  - Firestorms actually create their own local weather and are virtually impossible to extinguish.

Floods

- Flooding is a widespread problem in the United States that accounts for three-quarters of all presidential disaster declarations (PDDs).
- There are seven different types of flooding that are widely recognized.
- Flood risk areas in the United States are defined by the 100-year flood, an event that is expected to have a 100-year recurrence interval and, thus, a 1% chance of occurrence in any given year.
- Three types of automated devices detect imminent flooding: radar, rain gages, and stream gages.

Storm Surge

- Storm surge is increased height of a body of water that exceeds the normal tide, most commonly associated with hurricanes.
- Tsunamis are sea waves that are generated by undersea earthquakes, but volcanic eruptions or landslides can also cause tsunamis. Tsunamis are rare events. Over the course of a century, 15,000 earthquakes generated only 124 tsunamis.
- Evacuation to higher ground is the most effective method of protection. People must evacuate a long distance if they live on low-lying coasts.

Geophysical Hazards

- Threats from volcanoes include lightweight gases and ash that are blasted high into the air and heavier lava and mud that travel downslope.
- The impacts of volcanic eruption tend to be strongly directional because ash fall and gases disperse downwind.
- Within an earthquake's impact area, the primary threats are ground shaking, surface faulting, and ground failure.
- Protective measures can be understood by the common observation that "earthquakes don't kill people, falling buildings kill people."

Technological Hazards

- Hazardous materials (also known as hazmat) are defined as substances that are "capable of posing unreasonable risk to health, safety, and property" (Hazardous Material Definitions—Defined Term, https://definedterm.com/hazardous_material).
- Hazmat shipments result in an average of 280 liquid spills or gaseous releases per year, the majority occurring in transport.
  - 81% take place on the highway and 15% are in rail transportation.

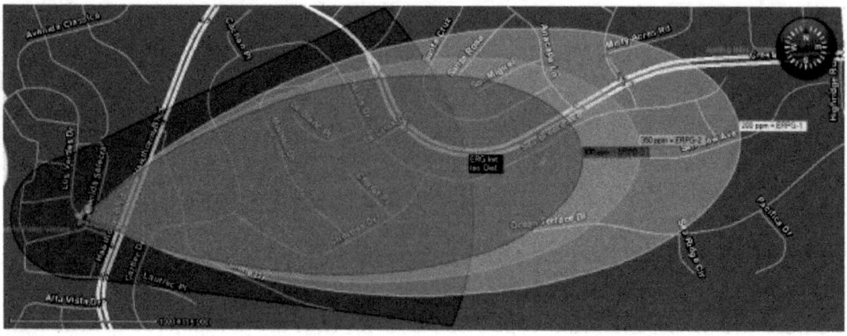

FIGURE 2.4 Plume model.

- Hazmat is produced, stored, or used in diverse locations, including the following:
  - Facilities such as petrochemical and manufacturing plants
  - Warehouses (e.g., agricultural fertilizers and pesticides)
  - Water treatment plants (chlorine is used to purify the water)
  - Breweries (ammonia is used as a refrigerant)
- Hazmat chemicals have three important temperatures—boiling point, flash point, and ignition temperature—that affect potential for fires.
- Toxic chemicals released from a storage container will produce an invisible cloud similar to the smoke plume that travels downwind from a campfire (Figure 2.4).
- Airborne dispersion of toxic industrial chemicals can produce serious inhalation exposures at distances up to 10 miles.
- Wind speed, wind direction, and atmospheric stability class affect the dispersion of a chemical.

Toxic Chemicals

- Terrorists might attempt to use weaponized toxic agents, which are toxic chemicals that require smaller doses to achieve disability or death.
- Capabilities needed to respond effectively to a toxic chemical attack are the same as those needed for an industrial accident. Few communities have these capabilities.
- In the case of radiological releases, even a late evacuation is better than home shelter; large building shelter is better than late evacuation; and early evacuation is best of all. The decrease in risk as a function of distance is quite large.

Biological Hazards

- Most biological agents that might be used in deliberate terrorist attacks also exist as natural hazards.
- Terrorists can spread disease by contaminating the food or water supply, or they can disperse biological agents in an aerosol cloud to cause inhalation exposure.

- It can be difficult to detect a biological agent infection because symptoms resemble the common cold and flu.
- If there is an outbreak, there are two actions to take: isolate and quarantine.

Summary

- As this section illustrates, each type of hazard has a different cause and effect. Each hazard also has a distinct set of characteristics and potential dangers. You must understand what to do to prepare for each type of hazard, inform others of the hazard, and aid others in the event that a hazard threatens your community.

Key Terms

- 100-year flood
- Earthquake
- Explosives
- Firestorms
- Flood
- Hazmat
- Hurricane
- Interface fires
- Natural hazards
- Severe storms
- Storm surge
- Technological hazards
- Tornadoes
- Tsunamis
- Wild land fires

HAZARD AND DISASTER CLASSIFICATION

Major Categories

- Natural hazards
- Anthropogenic nonintentional
- Anthropogenic intentional

Earth Hazardous to Your Health

- 516 active volcanoes, eruption every 15 days (average)
- 2,000 tremors daily
- Two significant earthquakes daily, severe damage 15–20 times annually
- 1,800 thunderstorms at any given time

Still Hazardous

- Lightning strikes 100 times per second.
- Late summer, an average of five hurricanes developing
- Four tornadoes per day or 600–1,000 annually
- 11 blizzards annually in the United States

Categories of Natural Hazards

- Atmospheric (meteorological)
- Geological (earth)
- Hydrological (water)
- Extraterrestrial
- Biological

Atmospheric-Sourced Processes

- Tropical cyclones
- Thunderstorms
- Tornadoes
- Lightning
- Hailstorms
- Windstorms
- Ice storms
- Snowstorms
- Blizzards
- Cold waves
- Heat waves
- Avalanches
- Fog
- Frost

Geological-Sourced Processes

- Earthquakes
- Volcanoes
- Tsunami
- Landslides
- Subsidence
- Mudflows
- Sinkholes

Hydrological-Sourced Processes

- Floods
- Droughts
- Wildfires

Extraterrestrial Processes

- Meteorites
- Asteroids
- Solar Flares
- Space Debris

Biological Processes

- Diseases
- Epidemics
- Pandemics
- Overpopulation
- Famine

Anthropogenic Nonintentional

- Technological
- Hazardous Materials
- Environmental
- Industrial
- Mining
- Nuclear
- Transportation
- Structural
- Industrial
- Mining

Technological

- Acts of people
- Technological systems that fail because of complexities and human fallibility (accidents)

Nuclear

- Power plants
- Industrial use
- Medical use

Transportation

- Aviation
- Highways
- Railroads
- Maritime

Structural

- Fires
- Collapse

Anthropogenic Intentional Hazards

- Mass shootings
- Civil disobedience
- Terrorism
- Weapons of mass destruction (WMD)

Mass Shootings

- School shootings
- Workplace violence
- Hate crimes
- Public shooting

Civil Disobedience

- Labor riots
- Race riots
- Political riots

Terrorism

- State/state sponsored
- International nonstate
- Domestic

WMD CBRNE (Chemical, Biological, Radiological, Nuclear, and High-Yield Explosives)

- Chemical
- Biological
- Nuclear/radiological
- Explosives

HAZARD, VULNERABILITY, AND RISK ANALYSIS

Community Vulnerability

- Communities can engage in three types of emergency management interventions to ameliorate disaster impacts.
  - Physical impacts can be reduced by hazard mitigation practices and emergency preparedness practices.
  - Social impacts can be reduced by recovery preparedness practices.

Preimpact Conditions

- Three pre-impact conditions must be considered:
  - Hazard exposure (living, working, or being in places that can be affected by hazard impacts)
  - Physical vulnerability (human, agricultural, or structural susceptibility to damage or injury from disasters)
  - Social vulnerability (lack of psychological, social, economic, and political resources to cope with disaster impacts)

Event-Specific Conditions

- Three event-specific conditions affect a disaster's impact on the community:
  - Hazard event characteristics
    - Each hazard has six significant characteristics.
  - Improvised disaster responses
    - When existing organizations seem incapable of meeting needs, they expand with new members or extend with new tasks, or new organizations emerge.
  - Improvised disaster recovery
    - Improvised recovery tasks may include damage assessment, debris clearance, reconstruction of infrastructure (e.g., electric power, fuel, water, wastewater, telecommunications, and transportation networks), and reconstruction of buildings in the residential, commercial, and industrial sectors.

Social Impacts

- Long-term social effects of disasters tend to be minimal in the United States.
  - Psychosocial impacts are usually mild and brief.
  - The largest demographic impacts of disasters are due to immigration and emigration.
  - Economic impact losses are difficult to measure because not all information is recorded.
  - Political impacts can create conflict as one group's attempts to solve its problems create problems for other groups.

Interventions

- There are three sets of actions that can reduce losses.
  - Hazard mitigation practices do not require people to take action when a disaster strikes. These practices usually involve physical preventive measures.
  - Emergency preparedness practices are preimpact actions that provide resources to support active responses at the time of hazard impact.

- Recovery preparedness practices are preimpact actions that can reduce the time it takes to recover after a disaster. Recovery from a major disaster takes much longer and involves much more conflict than people expect, and it is faster and more effective when it is based on a plan that has been developed before a disaster strikes.

Hazard and Vulnerability Analyses

- Mapping natural hazard exposure
  - One source is the set of maps contained in FEMA's Multi-Hazard Identification and Risk Assessment (Figure 2.5).
- Mapping hazmat exposure
  - To assess the risks, an inventory of facilities that use chemicals as well as vulnerable zones (VZs) that are likely to be affected by chemical releases can be created.
- Mapping exposure to secondary hazards
  - One way of identifying areas exposed to multiple hazards is to use a geographical information system (GIS) to overlay the areas subject to the different hazards.

Physical Vulnerability Assessment

- Buildings can be vulnerable to environmental hazards because of inadequate designs, inadequate construction materials, or both. It is very important to identify facilities that have special needs.

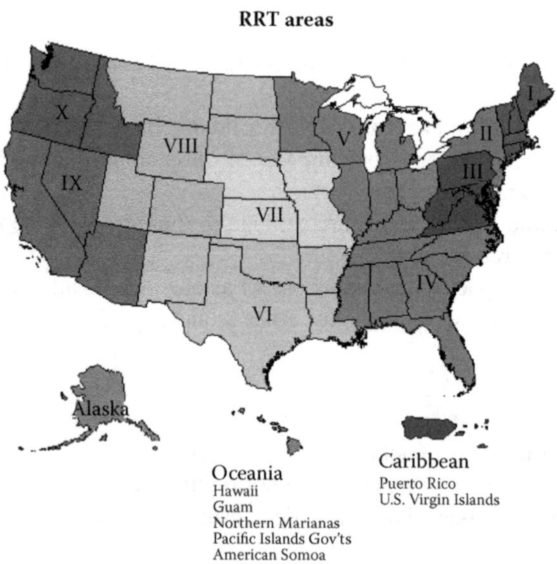

FIGURE 2.5 FEMA regions.

- To limit human vulnerability to inhalation exposure, buildings should have vapor barriers in the walls and ceilings for sheltering in-place.
- To limit human vulnerability to radiological materials, dense building materials such concrete, brick, and stone provide the best protection.
- Note that the emergency manager's job rarely involves assessing agricultural vulnerability or determining how crops and livestock will react to hazards.

Social Vulnerability Assessment

- The groups most likely to have high psychosocial vulnerability are the very young (who have limited verbal skills), the very old (who tend to be more isolated), and people with preexisting psychological problems.
- Households have most of their wealth in their homes, which are vulnerable to disaster impact.
- Households with high social vulnerability often occupy the areas most likely to be hit by a disaster and also occupy the oldest, most poorly maintained buildings.
    - Data can be combined on hazard exposure, structural lifeline vulnerability, and social vulnerability in a vulnerability hotspot analysis.

Vulnerability Analysis

- HAZUS-MH is a computer program that predicts losses from earthquakes, floods, and hurricane winds. It features three levels of analysis (Hazus Overview, https://www.fema.gov/hazus-mh-overview):
    - Level 1 is an initial screen that identifies the communities at highest risk.
    - Level 2 takes refined data and hazard maps to produce more accurate estimates.
    - Level 3 uses community-specific information to produce the most accurate loss estimates.
- Federal and state agencies have generated many hazard analysis documents that are available on various Web sites (e.g., Social Impacts of Disaster Training, fema.gov/hiedu/docs/fem/chapter).
    - This allows users to access information that might take those months to obtain if they were to request paper copy.

Assessing Risk

- FEMA's Multi-Hazard Identification and Risk Assessment program, as well as information from state and federal agency websites, can identify the hazards to which a community is exposed. This is a good start, but it is not enough.
- You need enough HVA data to decide how to allocate resources among hazard mitigation, emergency response preparedness, and disaster recovery preparedness.

- You also need HVA data to lobby for more money. Emergency management must compete with other community needs such as education, health care, and transportation. In this case, the HVA should be specific enough for you to persuade others that an increase in funding will benefit the community.

Summary

- As an emergency manager, you must be aware of the factors that make a community vulnerable to a disaster, identify key groups of people that are most likely to be affected, and understand the level of impact that a disaster will have on a community. With these details in mind, you can determine how to best protect your community using mitigation practices and emergency response practices.

Key Terms

- Agricultural vulnerability
- Emergency preparedness practices
- Hazard exposure
- Hazard mitigation practices
- HAZUS-MH
- Human vulnerability
- Multi-Hazard Identification and Risk Assessment
- Physical vulnerability
- Social vulnerability
- Vulnerable zone (VZ)

HAZARD MITIGATION

Hazard Mitigation

- Hazard mitigation is most effective when it takes place before disasters.
- The federal government cannot intervene directly in local land-use or building construction practices.
  - The federal government wants to change these practices because it pays for much of the high cost of disaster recovery.
- States have encouraged local governments to include hazard mitigation objectives in their everyday investment policies to reduce community hazard vulnerability.

Mitigation Strategies

- Hazard source control does not work for most natural hazards, with some exceptions.
- The four major types of flood control works are as follows:
  - Stream channelization

    - Dams
    - Levees
    - Floodwalls
- Levees have a number of design, construction, and maintenance problems.
- Floodwalls are built of strong materials, such as concrete, and are more expensive than levees, but they are also stronger.
- Landslide controls are designed to reduce shear stress, increase shear resistance, or a combination of the two.
- Industrial hazard controls are used to confine hazardous materials flows.
    - Dikes can be constructed around storage tanks to confine any liquid releases that might occur.
    - Such protection works are especially common around petroleum storage tanks.

Land-Use Practices

- The local government can influence land-use practices through the use of risk communication, incentives, and sanctions.
- Local governments have a variety of land-use management practices available to them that can be used to reduce hazard exposure.
- Some states require local governments to develop comprehensive plans that include land-use elements and environmental hazards elements.
- Tools available to local governments for hazard mitigation are zoning, subdivision regulations, capital improvements programs, acquisition of property or development rights, and fiscal policies.
- One advantage of a comprehensive plan is the increased likelihood that hazard mitigation measures will be area-wide rather than site-specific.
- A mitigation plan should be based on the HVA.
- Zoning can keep residential and commercial property away from earthquake fault lines, floodplains, explosions, and/or hazmat releases.

Building Construction Practices

- Property owners can change their construction practices voluntarily because of risk communication or incentives or can change involuntarily because of building code requirements.
- There are four major methods for providing protection from hydrological hazards.
    - The most cost-effective method is to raise a house so its lowest floor is elevated above flood level.
    - Dry flood proofing seals the structure so floodwater cannot enter. Walls are sealed with an impermeable coating.
    - Wet flood proofing allows water to enter empty portions of the structure during flooding.

    - Equipment is moved to a higher location or protected in place by a floodwall.
- Relocation involves moving the structure to higher ground out of the floodplain.

Structural Protection

- For wildfire hazards, a structure should be built with its exterior walls or roof constructed of nonflammable materials.
- For wind hazards, the best solution is to install stronger-than-normal connections among the foundation, walls, rafters or roof trusses, and roof decking.
- Many of the basic principles for protecting buildings from wind hazards also apply to seismic protection.
    - T-shaped, L-shaped, and U-shaped buildings are much more vulnerable than rectangular buildings.
    - Buildings constructed from masonry respond poorly to earthquakes because they are too rigid.
- Buildings can provide protection from inhalation of hazmat.
    - In-place sheltering is more effective than most people believe.
- Local building codes are based on model building codes established by nongovernmental organizations (NGOs).
    - Code adoption is only the first step in reducing vulnerability. The second step is to train building contractors and construction workers.
- For most hazards, protecting buildings from damage also protects the buildings' contents from harm.

Hazard Mitigation Measures

- There is an array of mitigation measures that can be effective for a variety of hazards.
- The federal approach to reducing toxic chemical hazards focuses on local emergency planning and community right-to-know (RTK) provisions.
- Local governments are expected to develop standards on hazardous facility siting in their land-use plans.

Natural Hazards

- Local governments can devise hazard mitigation programs. Such programs should emphasize the following:
    - Preventive actions must be decided at the local level.
    - Private sector participation is vital.
    - Long-term efforts and investments in prevention measures are essential.
- There is a connection between disaster reduction and sustainable development.
    - It is important to realize that development is not the same as growth.

Summary

- This section looked at hazard mitigation strategies and mitigation measures. It explained how building construction practices, land-use practices, and other mitigation legalities affect mitigation in communities. It is your responsibility to also know how to manage the risks of natural and technological hazards by reducing them.

Key Terms

- Channelization
- Floodwalls
- Hazard mitigation
- Industrial hazard controls
- Land-use practices
- Levees
- Right-to-know (RTK) provisions
- Sustainable development

DISASTER MYTHS, DISASTER DEMANDS, AND CITIZEN EMERGENCY RESPONSE

Response Myths

- People act in what they believe is their best interest, given their limited understanding of a situation.
- Emergency managers make poor decisions if they believe response myths.
    - The myth that people will panic becomes a reason to withhold information about a threat.

Shock and Panic

- Shock appears most frequently in sudden events involving widespread destruction, traumatic injuries, or death.
    - When symptoms do appear, few people are affected.
    - Shock lasts for a maximum of a few hours or days.
- The most stubborn myth about human response to disasters is that there is widespread panic.
    - When panic flight is observed, it involves a relatively small proportion of the people exposed.

Victim Response

- One common misconception is that there is a single "public" or "population."
    - In reality, population segments differ in their willingness and ability to evacuate.

- People in special facilities such as schools, hospitals, nursing homes, and jails have different patterns of warning and evacuation than residents and transients.

Integrative Responses

- The therapeutic community response is related to convergence.
  - Although convergence provides resources, it can also hamper response.
- A second aspect of the positive social response is sympathetic behavior from the public.
  - This promotes positive psychological outcomes for disaster victims; however, it is short-lived.

Warnings

- The effectiveness of a given warning method varies with the types of activities in which people are engaged.
  - Available warning methods are face-to-face, route alert, siren, commercial radio and television, tone alert radio, telephones, and newspapers.
- There is substantial variation across communities in the relative importance of different warning sources to different ethnic groups.
- People also use their own preexisting beliefs about appropriate protective actions.

Evacuation

- Evacuation trip generation refers to the number and location of vehicles evacuating from a risk area.
  - Evacuations initiated during daytime hours should include time for travel from work or school to home.
- As much as one-third or more of the households in some cities are dependent on public transportation (Public Transportation's Demographic Divide, www.governing.com/topics/transportation...public-transportation).
- Warning compliance refers to the percentage of people warned to evacuate who actually do so.
- Spontaneous evacuation refers to the percentage of people who were not warned to evacuate but do so anyway.
- When considering departure timing, you must consider two distinct population groups: residents and transients.
- It is important to estimate the percentage of early evacuating households (those who evacuate before an official warning) and the distribution of departure times for households leaving after an official warning is issued.
- There are four factors defining evacuees' destination/route choice:
  - Ultimate evacuation destination
  - Proximate destination (where the evacuees leave the risk area)

- Route choice
- Primary evacuation route utilization (the percentage of evacuating vehicles using official evacuation routes)

Search and Rescue (Disaster Research Center, https://www.drc.udel.edu)

- Forty-six percent of casualties reach hospitals in their own vehicles or in the vehicles of peers or bystanders.
- Evacuees prefer to avoid mass care facilities, with only 14.7% of evacuees going to shelters across all disasters.
- The Disaster Research Center's studies of hundreds of emergencies showed no evidence of role abandonment.
  - Hurricane Katrina was an exception.
- Volunteers must be placed into groups that perform tasks within the scope of their abilities.
- Disaster response organizations can be placed into four different categories (social associations, professional associations, industry associations, and hazards practitioners) Disaster Relief Agencies and Nongovernment Organizations www.disastercenter.com/agency.htm.

Household Behavior

- People threatened by disaster have multiple sources of information, and none is considered completely credible.
- Fear is rarely so overwhelming that it prevents people from responding, but it does impair people's ability to reason through complex, unfamiliar problems. To help people overcome fear, be sure to:
  - Provide clear, direct, relevant information about the hazard agent.
  - Inform people of the hazard's potential personal consequences.
  - Let people know what the authorities are doing to protect them.
  - Tell people how they can receive additional information.
- Official warning messages must include recommended protective actions.
- Without recommended protective actions, people take the most appropriate actions they already know or that they are told about by peers or the media (even if the action is incorrect for the hazard).
- The level of compliance with a warning is contingent upon the information source, message content, receiver characteristics, and situational characteristics such as threat familiarity and urgency for response.
- Some of those who are not at risk will also comply with protective action recommendations (spontaneous evacuation).

Stress and Health

- There may be cases of posttraumatic stress disorder among some population segments.
- Even in the apparent absence of psychological symptoms, victims and nonvictims have developed physical health problems following disasters.

- You should assume that many risk area residents have disaster-relevant competencies.
- Recognize the differences among households and, especially, avoid the error of assuming other households are just like yours.

Summary

- Because you can predict some behaviors, it's important you develop plans to respond to those behaviors. As emergency managers, you have to disseminate information to those who seek it, and you have to ensure that you've given appropriate warnings, solutions for protection, and even plans for organizing the people who show up to help after a disaster.

Key Terms

- Evacuation trip generation
- Panic

PREPAREDNESS FOR EMERGENCY RESPONSE AND DISASTER RECOVERY

Emergency Planning Principles

- Emergency planners should anticipate both active and passive resistance to the planning process.
- Preimpact planning should
  - Address all the community's hazards.
  - Elicit participation, commitment, and agreement among all response organizations.
  - Be based upon accurate assumptions about the threat, typical human behavior in disasters, and likely support from external sources such as state and federal agencies.
- EOPs should identify the types of emergency response actions that are most likely to be appropriate but encourage improvisation.
- Emergency planning should address the linkage of emergency response to disaster recovery and hazard mitigation.
- Preimpact planning should provide for training and evaluating the emergency response organization at all levels.
- Emergency planning should be recognized as a continuing process.

Response Functions

- Basic emergency response functions include emergency assessment and hazard operations:
  - Emergency assessment is the process of detecting a threat, predicting its potential impact, and determining how to respond.

    - Population monitoring and assessment identify the size of the population at risk and are important if the number of people varies over time.
  - Hazard operations involves taking actions to limit the magnitude of disaster impacts.
    - Plan in advance to develop procedures for contents protection practices. For example, locate sources of plywood for protecting windows from wind damage.
- Other basic emergency response functions include population protection and incident management:
  - Population protection involves taking protective actions to minimize the number of casualties.
    - Be aware that population protection tasks may require special equipment and procedures.
  - Incident management involves mobilizing and directing resources to respond to an emergency.
    - Develop procedures to ensure that key personnel are notified quickly.
    - Establish a space that will be used as the EOC.

Organizational Structures

- Organizational structures should
  - Form the basis of a larger structure to deal with disasters.
  - Be flexible.
- The ICS has seven basic principles (National Incident Management System [NIMS], https://training.fema.gov/nims).
- ICS was changed so that it could be used readily in both small and large events; the result was called the Incident Management System (IMS).
- The major advantage of IMS is that it makes all resources available for every incident.

Incident Management System

- The fundamental principle of IMS is that there must always be one and only one incident commander at every incident scene.
- Agencies select and train their staff to perform all the duties associated with specific standardized positions.
- Basic principles of IMS are easy to grasp, while more advanced concepts provide a sophisticated method of allocating responsibility for response strategy, tactics, and tasks.

IMS Implementation

- The goal of an articulated command is to spread functions to specialists.
- Most jurisdictions provide for the command staff to be supported by an on-scene public information officer and a police liaison.

- The advantage of IMS over the earlier ICS is that it provides for a better accounting of the activities that must be performed away from the incident scene.
- IMS is a flexible structure with close linkage between emergency plans and operations.
- IMS can only be executed effectively if emergency responders are thoroughly trained before incidents occur.

Emergency Operations Centers

- An EOC is a facility located in a safe area that provides support to responders at the scene of an incident.
  - An EOC is important because resources are often widely dispersed throughout a jurisdiction.
  - EOC staff includes both those who have technical knowledge and policymakers.
  - EOCs are used to provide close coordination among organizations at all levels.
- There are eight EOC design tasks: (1) assume position responsibilities, (2) common operating picture, (3) EOC processes and support systems, (4) leadership and accountability, (5) work group supervision, (6) operational requirements, (7) meetings, EOC briefings and debriefings, and (8) end of operational period/transition/demobilization.

Organizational Structures

- Three of the most important organizational structures include the following:
  - Metropolitan Medical Response System (MMRS) (MMRS Program, www.dhses.ny.gov/grants/hsgp/FY09_MMRS_Program_Guidance.pdf)
  - Urban Areas Security Initiative (UASI) (Homeland Security, www.dhs.gov)
  - NIMS (National Incident Management System, NIMS-FEMA.gov, https://www.fema.gov/national-incident-management-system)

Metropolitan Medical Response System

- MMRS tends to be concentrated in high population density areas and other areas that are terrorist targets.
  - Funding is provided directly to cities.
  - MMRS imposes a comprehensive emergency management process on recipient cities.

Urban Areas Security Initiative

- UASI's purpose is to prevent, respond to, and recover from acts of terrorism.
  - UASI requires a strategic plan for terrorist attacks anywhere in an urban area.

  - The program provides substantial funding for local needs.
  - UASI allows local choice in planning, administration, and funding.
  - There are concerns that under this program, federal authorities tightly define authorized expenditures within a budget category.

National Incident Management System

- NIMS was created in response to the terrorist attacks of 9/11.
  - NIMS is a standardized system for managing emergency preparedness and emergency response that builds on the ICS/IMS framework.
  - NIMS is broader than IMS because it addresses emergency preparedness and emergency response.
  - One concern many emergency managers and responders have regarding NIMS is the level of detail in which processes and protocols are specified.
  - NIMS requires significant commitments to training, drills, and exercises.

EOP Components

- EOPs should have four basic components:
  - Basic plan: The provisions of the basic plan are captured under seven separate headings.
  - Functional annex: Each annex should describe how the emergency response organization will perform a function needed to respond to disaster demands.
  - Hazard-specific appendixes: These provide information about the ways in which the response to a particular hazard agent differs from the standard response.
  - Standard operating procedures and checklists: These list the steps that individuals and organizations will take to perform specific emergency response tasks.

Summary

- As an emergency responder, you strive to prevent casualties and damage in a disaster situation. As evidenced from recent natural disasters such as Hurricanes Katrina and Rita and terrorist attacks such as those on 9/11, emergency managers must develop a state of readiness to respond to extreme events threatening their communities.

Key Terms

- Emergency assessment
- Functional annex

ORGANIZATIONAL EMERGENCY RESPONSE

Emergency Assessment

- Emergency assessment activities in the response phase are directed toward intelligence.
- Threat detection includes the following:
  - Recognizing that a threat exists.
  - Assessing the threat's magnitude and location and the timing of impact.
  - Determining how to respond.
- An emergency classification system organizes a large number of potential incidents into a small set of categories.
  - Categories link the threat assessment to the level of activation of the responding organization.

Hazard Monitoring

- The hazard agent must be tracked over time to determine whether the threat is changing in its likelihood, magnitude, immediacy, or location of impact or whether there are changing environmental conditions.
- For hazmat releases, responders should do the following before the appropriate monitoring equipment arrives:
  - Establish a perimeter around the scene.
  - Prevent entry into the "hot zone."
  - Decontaminate anyone who has been exposed.
  - Maintain medical observation and isolation until released by a competent medical authority.

Monitoring and Assessment

- Census data documents the size and composition of the community's permanent population.
  - It's important to break the population down by age and ethnic composition.
  - Estimates of workers, tourists, and other transients should also be added.
- Damage assessment focuses on measuring the impacts of a disaster on public and private property.
  - This is a continuing process that begins during emergency response.

Hazard Operations

- Hazard operation actions are implemented only when the need arises. These actions must be able to be implemented rapidly.

  - Hazard operations actions can be grouped into the same categories as permanent hazard mitigation measures.
  - Hazard operations measures are not feasible for hazards such as tornadoes and earthquakes in which there is insufficient forewarning.
- Standard operating procedures should contain rules that define the conditions under which each hazard operations action should be used or avoided.

Population Protection

- Information collected through the emergency assessment function forms the basis for the population protection function.
- The appropriate protective strategy varies based on the following factors:
  - Type of environmental threat
  - Certainty of occurrence
  - Severity of impact
  - Immediacy of impact
  - Duration of impact
- During an emergency, specific warnings should be disseminated by expert and credible sources.
  - These warnings should describe the threat in terms of its location, severity, and expected time of impact.

The warnings should also recommend appropriate protective actions and indicate how to obtain additional information.

Protective Actions

- People might need the following elements to implement protective actions:
  - Money
  - Knowledge and physical skill
  - Facilities and vehicles
  - Tools and equipment
  - Time and energy
  - Social cooperation
- Three population segments will need help in an evacuation.

Population Protection

- To establish impact zone access control and security, you should take the following steps:
  - Set up traffic control points on major routes at the perimeter of the impact area to prevent people from entering without authorization.
  - Provide security patrols within the impact area.

- For search and rescue operations, you should directly address the need for heavy rescue. This includes the following tasks:
    - Assigning a lead agency.
    - Maintaining a list of available heavy rescue equipment.
    - Establishing decision criteria for prioritizing buildings in the event of multiple collapses.
    - Creating a protocol for quickly obtaining services of victim location specialists.

Care of Victims

- Emergency authorities provide short-term support for victims in the form of food, accommodations, and limited medical care.
- Evacuee registration provides a link to the population monitoring function, especially accountability and casualty assessment.
- Experience has shown that not allowing pets discourages needy families from using shelters.

Emergency Medical Care

- Medical care for victims of major disasters is provided by three components of the emergency response:
    - Emergency personnel in the field
    - The network of local hospitals
    - The National Disaster Medical System, which is a system of military aircraft equipped to sustain treatment
- Triage is used to sort victims into categories so that emergency responders can make the most effective use of medical treatment to save the maximum number of lives.
- In the United States, the county ME's office typically performs the morgue function.

Exposure Control

- Hazard exposure control seeks to reduce exposure to a level that is as low as reasonably achievable.
- Emergency personnel can achieve the as low as reasonably achievable (ALARA) objective in three ways:
    - Minimizing the amount of time spent in hazardous areas (time).
    - Staying as far away from hazard sources as necessary (distance).
    - Insulating themselves from the hazard (shielding).
- Many communities use hazmat response teams to test for, abate, and monitor chemical, biological, and radiological (CBR) contamination.
- A CBR contaminated area might also be a crime scene and may therefore involve FBI participation.

Incident Management

- The concept of operations is a summary statement of what emergency functions are to be performed and how they are to be accomplished.
- Notification to the jurisdictional authorities comes from different sources, depending on the nature of the threat.
    - The notification process differs among different types of hazards.
- EOCs are the hub of emergency information processing within the jurisdiction.
    - The need for organizations to be represented in the EOC depends on the nature of the threat as well as the network of governmental resources needed to respond to that threat.
- The communications team collects, displays, and records data on environmental conditions, casualties, and damage to property and the environment.
- Communication logs should record who called into or out of the EOC, when the call took place, who participated in the call, and the content of the call.
- Analysis/planning teams assess the current status of the situation and project its future status.
- The location from which public information is given depends upon the size of the incident and upon whether emergency operations are taking place at a defined incident scene.
    - Incident managers should designate a broadly knowledgeable chief spokesperson who can call upon specialists to respond to specific questions.
- Administrative and logistical support is handled at the incident command post during minor events. However, it is transferred to the EOC during a community-wide disaster.

Summary

- The EOP and procedures developed during the emergency preparedness phase affect the actions you take during emergency response. However, no emergency can ever be predicted with accuracy; therefore, it is important that you learn to improvise effectively when the need arises. Focus on the four basic functions: emergency assessment, hazard operations, population protection, and incident management. Understand what tasks are associated with each of these basic functions.

Key Terms

- Concept of operations
- Damage assessment
- Emergency classification system

DISASTER RECOVERY

Communities

- A community is a specific geographic area (town, city, or county) with a government.
    - Two additional elements of community are psychological ties and social interaction.
- If a disaster causes members of a community's households to change their normal patterns of daily activity, this can cause psychological and economic distress and social conflict.
- A community's industries that produce exports should receive immediate attention in the aftermath of a disaster to stimulate economic activity.

Recovery Process

- Recovery begins when the emergency has been stabilized so there is no longer a threat to life and property.
- Disaster recovery is both physical and social.
- Restoring the community to its previous status can reproduce the hazard vulnerability that led to the disaster.
- A disaster-resilient community learns how to use a disaster as a focusing event that changes people's behavior.
- Developing preimpact plans for disaster recovery is important because there will not be much time for recovery planning after disaster strikes.

Household Recovery

- Households typically pass through four stages of housing recovery following a disaster: emergency shelter, temporary shelter, temporary housing, and permanent housing.
- Households lacking adequate insurance coverage must use other strategies for recovery, such as the following:
    - Obtaining Small Business Administration or commercial loans
    - Seeking FEMA or NGO grants
    - Withdrawing personal savings
    - Not replacing damaged items
- There are differences in the rate of economic recovery among ethnic groups.
- Few victims develop major psychological problems from disasters. Most people experience only mild distress.
- Work with local mental health agencies to ensure that an organized mental health referral system will be available for people in higher risk groups.
- Households can be characterized in terms of three modes of disaster recovery:
    - Autonomous recovery
    - Kinship recovery
    - Institutional recovery

Business Recovery

- During and after a disaster, disruption of infrastructure such as water/sewer, electric power, fuel, transportation, and telecommunications may force businesses to shut down.
- Small businesses are more vulnerable than large businesses for two important reasons:
  - They are more likely to be located in nonengineered buildings that become damaged.
  - They are less likely to have hazard management programs to reduce their physical vulnerability.
- Wholesale and retail businesses report significant sales losses following disasters. However, manufacturing and construction companies often show gains.

State and Federal Governments

- The burden of disaster recovery most frequently falls on local governments.
- In large events, PDDs open up a broad range of programs for relief and reconstruction.
  - The state plays a coordinating role between the federal government and local governments.
  - The federal government will influence state and local behavior during the recovery period.
- The lead recovery agency at the federal level is FEMA.
- The main types of programs providing recovery assistance are the individual assistance, infrastructure support, and hazard mitigation grant programs.

Local Government Recovery

- By planning for recovery before disaster strikes, a community can allocate resources more effectively and efficiently during and after a disaster.
- There are three major misconceptions about recovery:
  - Recovery can be improvised after the response is complete.
  - There is ample time to collect data and plan the recovery during emergency response.
  - The objective of recovery should be to restore the community to the conditions that existed before the disaster.
- A recovery/mitigation committee should work with the rest of the community to formulate a vision of the type of disaster recovery it intends to implement.

Recovery Operations Plan

- Disaster assessment should include both physical and social impact assessment.

- There are three types of damage assessment:
  - Rapid assessment
  - Preliminary damage assessment
  - Site assessment
- In preparing for damage assessment, local government staff should be trained in a common assessment procedure.
- The psychological, demographic, and economic impacts of disasters can be evaluated using a victims' needs assessment.

Short-Term Recovery

- The first benefit of disaster assessment is that it guides short-term recovery, which consists of the following eight tasks performed in the immediate aftermath of a disaster:
  - Impact area security and reentry
  - Temporary shelter/housing
  - Infrastructure restoration
  - Debris management
  - Emergency demolition
  - Repair permitting
  - Donations management
  - Disaster assistance

Long-Term Reconstruction

- A disaster opens a window of opportunity to change a community's hazard management policy.
- Over the long-term, a community should avoid increases in vulnerability by making changes in its land-use and building construction practices.
- During reconstruction, waterborne illnesses may be a problem if survivors drink from, wash food in, or bathe in water sources that have been contaminated.
- Note that natural disasters produce minimal mental health consequences.
- Hazmat spills are an increasing problem during natural disasters.
- In communities that are highly dependent on tourism, active promotion is needed to assure potential visitors that all facilities are back in operation following a disaster.

Recovery Management

- To manage disaster recovery, you need to perform many of the same types of tasks as during the incident management function of the emergency response phase.
- The recovery process typically involves the same tasks that agencies perform as part of their normal duties.
- The ROP should describe the procedure for providing public information.

- The recovery/mitigation committee needs to obtain legal authority for a wide range of short-term recovery actions.
- Special provisions are required to support the additional staff generated by obtaining mutual aid personnel from other jurisdictions and volunteer personnel such as architects and engineers used as building inspectors.

Summary

- Once disaster strikes, recovery takes center stage as many organizations and people come together to rebuild. Recovery involves households, businesses, and communities. Working together and understanding the obstacles can aid in the recovery process. It is important to utilize the state and federal governments and various charities to assist people and groups in restoring the community's normal patterns of social functioning.

Key Terms

- Community
- Emergency shelter
- Permanent housing
- Preliminary damage assessment
- Rapid assessment
- Site assessment
- Temporary housing
- Temporary shelter
- Victims' needs assessment

EVALUATION

Performance Appraisals

- Personnel performance appraisals serve four functions: development, reward, internal research, and legal protection.
- Appraisals should be conducted at least once a year.
- Performance should be rated based on data that meet three conditions:
  - The data must be available within the time period in which the appraisal is being conducted.
  - The data must be relevant to job performance.
  - The data must be comprehensive.
- When conducting an appraisal, ensure that both the short-term and long-term aspects of the job are evaluated.
- Human resource departments typically have performance appraisal criteria devised for all civil service jobs.
- There are nine typical performance appraisal categories.

- If you have an overall positive performance appraisal, do not dwell on the negative aspects of the appraisal. Unfortunately, dwelling on the negative is quite common.

Organizational Evaluation

- Work with the other members of the LEMA and the LEMC to set specific, measurable objectives that can be accomplished within the period of performance.
- Evaluating the performance of the LEMC is more complex than evaluating that of the LEMA, but it involves basically the same procedures.
- Your control over the allocation of resources in the LEMC is more limited than your control over the LEMA.

NFPA Standard 1600

- The National Fire Protection Association (NFPA) Standards Council established a Disaster Management Committee in 1991 that developed standards for preparedness, response, and recovery ("NFPA Publishes New Disaster and Emergency Management," www.facilities.net.com/emergencypreparedness/article/NFPA-Publishes).
- The committee adopted what it called a "total program approach."
  - The standard covers both public and private sector organizations.
  - Business continuity programs are included in the scope of the standard.
- Programs established under the standard must address 14 elements.
- The standard requires organizations to have the following components in place:
  - A documented emergency management program
  - An adequate administrative structure
  - An identified coordinator
  - An advisory committee
  - Procedures for evaluation
- The Hazard Mitigation element of the standard requires organizations to develop a strategy to eliminate hazards or limit their consequences.
- NFPA 1600 requires the specification of roles and responsibilities. Roles must be defined for external organizations that will participate in mitigation, preparedness, response, and recovery activities.
- The Direction, Control, and Coordination element specifically includes adopting an IMS and assigning functional responsibilities to specific entities within the organization.
- The Operations and Procedures element requires the organization to develop the procedures needed to respond to the hazards identified in the Hazard Identification, Risk Assessment, and Impact Analysis element.
- The Crisis Communications, Public Education, and Information element requires the entity to establish procedures for disseminating relevant information to the news media and the public during the pre-, trans-, and post-disaster phases of operation.

- NFPA 1600 is important to the emergency management profession for several reasons:
  - It was issued by a respected and established authority. The government and major emergency managers respect and understand the importance of NFPA standards.
  - It is extremely useful in program assessment. The standard provides a model that can be used in self assessment and also by external evaluators.
  - It can serve as a basis for planning to create a program or for planning to enhance an existing program so that it meets the standard.
  - It can be used as the authoritative basis for your position when making an argument to expand or change your program.

Capability Assessment for Readiness Program

- In 1997, FEMA and the National Emergency Management Association released the State Capability Assessment for Readiness (CAR) program (www.allhandsconsulting.com/.../capability-assessment-for-readiness-car).
- The CAR program describes a self-assessment process for SEMAs to do the following:
  - Evaluate their readiness to mitigate hazards
  - Prepare and respond to emergencies
  - Recover from disasters
- The CAR program consists of standards that begin, at the highest level, with 13 emergency management functions adapted from NFPA 1600.

Emergency Management Accreditation Program

- The Emergency Management Accreditation Program (EMAP) is closer to NFPA 1600 than CAR because it includes the requirements for program management (https://www.emap.org/.../the-emergency-management-standard).
  - EMAP was written specifically for state and local emergency management agencies.
  - The EMAP accreditation process is more elaborate than the CAR assessment. Once a jurisdiction submits an application, it has 18 months to conduct a self-assessment of its compliance with EMAP's 54 standards.
- The self-assessment portion of the process requires a proof of compliance record for each standard.
- If accredited, a jurisdiction is issued a certificate that is valid for five years.

Drill, Exercises, and Incidents

- Drills are used to test people, facilities, and equipment on tasks that are difficult, critical, and performed infrequently.

- A drill is conducted by a controller, the person who provides the information from the scenario. Drills are relatively simple, so the same person can also serve as the evaluator.
- Some tasks have a significant mental as well as physical component.
- A functional exercise differs from a drill by involving more people.
  - Functional exercises are more comprehensive than drills.
  - The scenario for a functional exercise is usually more complex because it involves more tasks, equipment, and people.
  - Functional exercises test people's ability to perform both task work and teamwork.
- Unlike drills, functional exercises cannot combine the roles of controller and evaluator. In fact, exercises sometimes require many controllers.
- A full-scale exercise simulates a community-wide disaster by testing multiple functions at the same time.
- The complexity of full-scale exercises requires thorough planning of the scenario as well as coordination among the many controllers and evaluators.
- Large exercises conducted for nuclear power plants can involve thousands of players and as many as 50 to 100 controllers and evaluators.
- Most full-scale exercises are unannounced.
- FEMA organizes the development of an emergency exercise into eight steps.
- Evaluations of performance in incidents are extremely informative because incidents are unscheduled.
- The disadvantage of using actual incidents for evaluations is that incidents are uncontrolled. Both the size of the event and the response functions that are tested are matters of chance. Incidents also have no controllers or evaluators.
- All three forms of exercises and incident responses benefit from an oral critique by the players, controllers, and evaluators.
- Critique results should be documented in a written report that includes an action plan. Be sure to write specific recommendations, assign responsibility for implementation, and schedule completion of each element of the action plan.

Training and Risk Communication

- The procedures for evaluating training and risk communication programs are different from the procedures for the previous types of evaluations.
- Training and risk communication programs are usually administered to many more people than you can afford to evaluate. Thus, it may be best to test a subset (called a sample) of individuals from the larger group that received the training or risk communication program (called the treatment group).
- The performance of this sample from the treatment group can then be compared to the performance of a group of people who did not receive the training or risk communication program (called the control group).

Summary

- Given how important emergency management is, you must hold people accountable for their job responsibilities in this area. This chapter shows you how to conduct your own performance appraisals and how to prepare drills and exercises to assess the performance of your emergency response organization.

Key Terms

- CAR program
- Controller
- Drills
- Evaluator
- Full-scale exercise
- Functional exercise
- Task work
- Teamwork

INTERNATIONAL EMERGENCY MANAGEMENT

Policy Variations

- In disaster management, the influence of the United Nations has contributed to the use of common models.
- Countries with high levels of exposure have been described as having "disaster cultures" (Disaster Risk, PreventionWeb.net, www.preventionweb.net/risk/disaster-risk).
- Emergency management is low on the priority list in poorer countries.
- The quality of emergency management in a country is related to the amount of internal and external resources available.

Economic Resources

- Hazard insurance availability varies from one country to the next.
- Availability of "specialized assets" is a factor affecting emergency management. These assets may include heavy equipment, trained urban search and rescue teams, hazmat capabilities, technical expertise such as GIS, and training facilities.
- The control of policies, programs, and resources at the national level limits the ability of local governments to mount a rapid emergency response.

Government Organization

- Too much emphasis on large disasters tends to lead to the over centralization of disaster management.
- Few countries have an adequate supply of well-trained emergency management professionals.

- The quality of a country's infrastructure and housing affects its level of disaster exposure.
- When members of the public are well informed and have strong beliefs in their rights, they are likely to demand competence from their government.

Military and Organizations

- The armed forces are involved in emergency management to some degree almost everywhere.
- The armed forces usually have more of the resources necessary for disaster response.
- Countries vary widely in their approach to foreign aid.
- Many international institutions are devoted to promoting improved emergency management practices.
- In addition to the United Nations, there are regional groups involved in emergency management.

Emergency Management in Brazil (https://training.fema.gov/hiedu/docs/fem/chapter)

- Brazil is a republic whose states have strong political powers.
- Emergency management in Brazil has developed over time to reflect the structure of the country's government and the country's hazards.
- Landslides are the most common cause of death from natural disasters in Brazil. Many factors contribute to this situation.
    - The Southern Hemisphere's summer rainy season is from December through March. During this time, emergency managers in the Serra do Mar activate a landslide monitoring system by entering the Observation stage.

New Zealand Restructuring

- New Zealand is exposed to floods, earthquakes, tsunamis, and cyclones.
- Local financial and environmental management plans enabled communities to integrate emergency management into land-use planning and development.
- A comprehensive approach was adopted that incorporated all hazards and all phases of emergency management.
- New Zealand has sought a balance between centralization and localization of emergency management.

India's Recovery

- As in many poor countries, emergency management in India has been primarily reactive, concentrating on relief after major disasters.
- The Maharashtra Emergency Earthquake Reconstruction Program focused on reconstruction and rehabilitation (India—Maharashtra Emergency Earthquake Rehabilitation documents.worldbank.org).

- Disaster management planning gave the state a unique level of expertise. In later disasters, other Indian states were able to draw upon this expertise.
- The Patanka project used a three-stage process that was developed and implemented over a two-year period ("Patanka Project—Community Recovery and Its Sustainability: Lessons from Gujarat Earthquake of India," https://ajem.infoservices.com.au/downloads/AJEM-18-02-03).
- The project team worked closely with local leaders to win community trust and develop strong local leadership.

Land-Use in Colombia

- Colombia is vulnerable to several hazards, including volcanoes, earthquakes, floods, and landslides.
- Colombia's Territorial Development Law includes provisions for natural hazard risk assessment, land-use planning, and urban development ("Organic Law of Territorial Organization in Colombia," https://wp.nyu.edu/mariamonicasalazartamayo/wp-content/uploads/). The law provides detailed directions on the elements to be included in land-use plans.
- Colombia's law is remarkable because its requirements are more stringent than those that many states in the United States place on local governments. This law is also interesting in the way it incorporates hazard mitigation directly into the fabric of local land-use planning.

Seveso Directives

- The Seveso Directives (Seveso Directives—Major Accident Hazards—Environment—European ec.europa.eu/environment/seveso. It also is a plan for international use with US involvement):
  - Incorporate many of the community RTK provisions set forth in SARA Title III.
  - Go further by requiring active dissemination of information to the public.
  - Use land-use planning to manage chemical hazards.
- The primary goal of the Seveso Directives is to prevent hazmat accidents. The secondary goal is to limit the health, safety, and environmental consequences of any accidents that do occur.
- Seveso II requires the operator of each hazmat facility to develop an on-site emergency response plan and supply it to local authorities. The local authorities must then use each facility's on-site plans to develop their own EOPs.

Chi-Chi Earthquake

- In 1999, a 7.6-magnitude earthquake hit Taiwan. Known as the Chi-Chi earthquake, it caused 2,400 deaths and 11,000 injuries requiring medical attention (Chi-Chi, Taiwan Earthquake Event Report, forms2.rms.com/rs/729-DJX565/images/eq_chi_chi_taiwan_eq.pdf).

- Local government lacked the resources and experience to respond to the disaster until the central government could overcome its lack of emergency assessment capability.
- The educational infrastructure of the affected area was severely damaged; 75% of the area's schools were closed.
- The government's social services were also stretched to the limit and beyond after the Chi-Chi earthquake.

Summary

- People are now much more aware of natural disasters than they were in the past. And, countries are able to collaborate more easily to provide assistance. Responding to natural disasters often brings the world together, as evidenced by the devastating tsunami of December 2004.

PROFESSIONAL ACCOUNTABILITY

Distinguishing Emergency Management

- Emergency managers have knowledge about a wide range of hazards and must know how to manage a community's vulnerability.
- Emergency responders directly respond to disasters, attacking the threat to reduce potential disaster losses.
- Public-sector emergency managers work for all levels of government—federal, state, and local.
- Private-sector emergency managers work for organizations such as chemical facilities, nuclear power plants, and railroads.
- The final distinction is among local, state, and federal emergency managers.
- For a city or town, emergency management rarely exists as a separate department. It is often located within a fire or police department.
- State and federal emergency managers have positions that are quite different from those of local emergency managers.
- FEMA works with other agencies, such as the Environmental Protection Agency, Coast Guard, and Department of Transportation.
  - Together, they develop programs and provide technical and financial assistance to LEMAs.
- The principal requirements are the willingness and ability to work with multidisciplinary teams.

Defining "Profession"

- A profession holds its members accountable to their peers for behavior that is relevant to the profession.
- Professions have membership rules to exclude unqualified people.
  - Membership rules usually relate to education and training requirements.
- A profession has an "evolving and agreed-upon body of knowledge."
- A defining feature of professions is that they have ethical standards.

Emergency Management as a Profession

- Most people would agree that emergency management is a profession.
- The current model for emergency managers is that of a career-oriented, college-educated professional who has acquired knowledge from the physical and social sciences.
- An emergency manager must possess many skills, including communication skills, organizational skills, and the ability to grasp the technical fundamentals of a range of threats.
- Emergency managers are generalists who know where to find and how to request the services of specialists.

Development and Ethics

- Emergency managers should participate in one or more professional associations.
- Continuing education credits are offered through a variety of colleges and universities, and FEMA's Emergency Management Institute also offers professional development opportunities.
- The International Association of Emergency Managers (IAEM) emphasizes ethics among its members and has adopted a three-part formal ethical code (iaem.com):
    - Respect people, laws, regulations, and fiscal resources.
    - Gain trust, act fairly, and be effective stewards of resources.
    - Embrace professionalism founded on education, safety, and protection of life and property.

Body of Knowledge

- Emergency management is an interdisciplinary profession, so you must always draw upon knowledge in the physical and social sciences.
- In addition to their other duties, emergency managers may want to conduct research.
- In asserting their credibility to the public, emergency managers are set apart from other professions by their body of knowledge.
- Organizations are attempting to explicitly define the emergency management body of knowledge. The FEMA Higher Education Project is contributing to this effort by examining how to develop an accreditation system for degree programs.

Emergency Management Certification

- When academic degrees are unavailable, certificates are especially important to ensure that you have received the necessary technical training.
- Business continuity planning offers a wide range of certificates.
- There are many new programs of unproven quality.

- The critical certification is that of Certified Emergency Manager (CEM), which is offered through the IAEM.
- The CEM is the only certification that assures competence in comprehensive emergency management and integrated emergency management systems.

Academic Programs

- The growth of academic degree programs to support a profession represents the maturing of that profession.
- Professional degree programs do the following:
  - They help you acquire principles and procedures from different theoretically organized disciplines.
  - They bring together practitioners and researchers.
- To serve the profession, degree programs must achieve some level of standardization. There must be an assurance that graduates of degree programs know the body of professional knowledge.
- FEMA has created a learning resource center and posted sample syllabi for a wide range of classes. The agency has also developed full college courses with instructor guides, readings, exercises, field trips, and student notes.

Legal Liability

- Legal liability applies more to organizations and government agencies than to individuals.
- It is important to note that emergency management statutes vary widely among the states, as do the emergency powers that are available to address disasters.
- Two areas of legal concern commonly arise:
  - The claim that, in responding to an emergency, government officials caused damage to persons or property.
  - The claim that a government's failure to plan for or respond to a disaster resulted in damage to persons or property.
- Three exceptions to immunity are important for emergency managers:
  - Claims for damages may not be brought in connection with the imposition of quarantine.
  - A person may not bring suit if federal agencies or employees can demonstrate that they exercised "due care" in carrying out a statute.
  - The "discretionary function exception" provides immunity for federal agencies and employees when a claim is based on the "exercise or performance or failure to exercise or perform" a discretionary action.
- Immunity is recognized unless it can be shown that some form of negligence exists.
- To mount a successful defense, a jurisdiction needs to document that it has a technically sound emergency plan.

Summary

- Emergency management is a challenging career that is maturing as a profession. There are education, training, and certification opportunities, as well as professional associations you can belong to that will help you in your career.

Key Terms

- Certification
- Emergency manager
- Emergency responder
- Profession

FUTURE DIRECTIONS IN EMERGENCY MANAGEMENT

Global Challenges

- Scientists agree that climate change is a fact and that the consequences of this change are likely to be serious.
- Climate change will increase the number of extreme events taking place across the globe.
  - It will increase the number of severe storms and floods.
  - It will lead to an increase in drought conditions across the plains of Africa, North America, and South America.
- The world's population may increase by as much as 50% in the next 50 years (www.un.org/press/en/2005/pop918.doc.htm).
  - The biggest increases will be mostly in the developing countries of Asia and Africa.
  - The population boom will create greater demands for food, water, and energy.
  - Shortages will bring about political instability and breeding grounds for terrorists.
- The concentration of population and wealth in urban areas will create the potential for a mega-disaster.
- Worldwide consumption of fresh water supply may reach 90% by 2030.
  - Two-thirds of the world's population will experience chronic shortages of safe drinking water by that time.
- Income and wealth are becoming increasingly concentrated in the hands of the wealthiest.
- Residents of developed countries are demanding that many technologies become safer.
  - Few of these people, however, are willing to pay higher taxes.

- Emergency managers must be prepared to "make do" with current levels of resources, respond quickly during and after disasters, and be patient with some people's unrealistic expectations.

Opportunities

- Policymakers need to understand the causes of hazards to manage them effectively.
- People are becoming aware that the state of our environment is a major factor in the occurrence of disasters.
- There are many rapidly developing technologies to support emergency management.
- You will be more effective if you can show that it is less expensive to mitigate hazards than to rebuild after a disaster.
  - The data you need, together with computer programs such as HAZUS, are becoming increasingly available.
- The greatest advance in technology has been the availability of GISs.
- Emergency response technologies have improved forecast and warning systems for a variety of different hazards.
- There are more ways to record and communicate information than ever before.
- There will continue to be significant advances in the communication technology used to warn residents about hazards.
- In disaster recovery, you can use many of the common technology tools to quickly assess damage and send the information back to the EOC.

National Challenges

- Much of the future increase in the U.S. population will occur in hazard-prone areas.
- Emergency managers need to promote smart growth that minimizes disaster losses.
- Property owners in disaster-prone areas are being subsidized by other taxpayers.
- There will continue to be an increase in the cultural and language diversity of the population.
- Rich communities will be able to adopt new technology, but poor communities may not have the money to do so.

Terrorist Threats

- Plans for responding to terrorist threats can include many of the procedures you expect to use for technological accidents.
- Terrorist threats also present new challenges, however.
  - Exotic chemicals such as sarin gas, "dirty bombs," and biohazards present different problems from technological accidents.
- An intelligent adversary creates some important information security problems.

- Terrorists can take advantage of predictable population protective responses to inflict even greater casualties in secondary attacks during mass evacuations.

National Challenges

- In the past, emergency management was a low priority on the government agenda. This changed dramatically after 9/11.
- It is important for emergency managers to build coalitions with other agencies, NGOs, and private-sector organizations.
- Legal liability is a major issue because each state has different rules regarding liability in an emergency response.
- There is a conflict between the goals of economic development and private property rights versus the goals of public safety and welfare.

Professional Challenges

- Many emergency managers have limited contact with land-use planners and public health departments.
- Cooperation between emergency managers and disaster researchers is not easy.
    - This situation seems to be changing for the better.
    - Professors in emergency management and related fields now have greater contact with emergency managers.
    - There are increasing numbers of research projects designed to solve practical problems.

Professional Opportunities

- Recent years have seen an increase in graduates holding postsecondary degrees in emergency management.
- The need for emergency managers to become involved in hazard mitigation has been recognized for many years.
- Preimpact recovery planning provides opportunities to work with land-use planners and building construction officials.
- Emergency managers' knowledge and skills in preparing communities for unexpected events makes them an invaluable consultant to senior administrators.
- Collaboration among jurisdictions within a region or between levels of government requires some degree of organizational standardization.

Summary

There are many challenges and opportunities facing emergency managers at the global, national, and local levels. It is essential that communication improves so that emergency managers can collaborate with other professionals for better preparedness.

3 Security Management

Security management is the identification of an organization's assets (including information assets), followed by the development, documentation, and implementation of policies and procedures for protecting these assets.

An organization uses such security management procedures as information classification, risk assessment, and risk analysis to identify threats, categories assets, and rate system vulnerabilities so that they can implement effective controls (Figure 3.1).

LOSS PREVENTION

Loss prevention focuses on what your critical assets are and how you are going to protect them. A key component to loss prevention is assessing the potential threats to the successful achievement of the goal. This must include the potential opportunities that further the object (why take the risk unless there's an upside?). Balance probability and impact determine and implement measures to minimize or eliminate those threats.

SECURITY RISK MANAGEMENT

Management of security risks applies the principles of risk management to the management of security threats. It consists of identifying threats (or risk causes), assessing the effectiveness of existing controls to face those threats, determining the risks' consequence(s), prioritizing the risks by rating the likelihood and impact, classifying the type of risk, and selecting an appropriate risk option or risk response (Figure 3.2).

TYPES OF SECURITY THREATS

External

Strategic: like competition and customer demand
Operational: regulation, suppliers, contract
Financial: Forex (FX) credit
Hazard: natural disaster, cyber, external criminal act
Compliance: new regulatory or legal requirements are introduced, or existing ones are changed, exposing the organization to a non-compliance risk if measures are not taken to ensure compliance

Internal

Strategic: R&D operational: systems and process (H&R, Payroll)
Financial: liquidity, cash flow

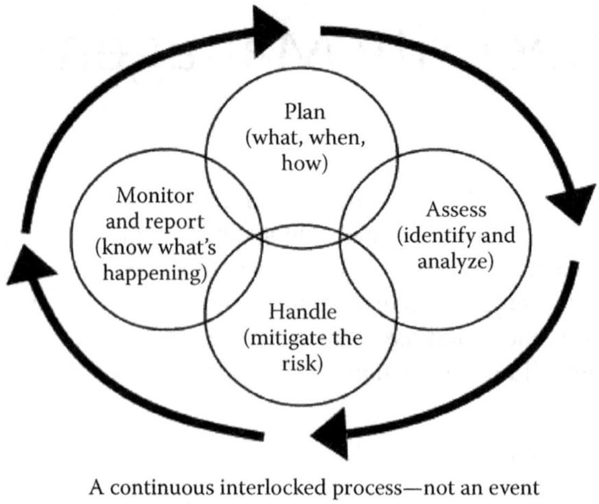

FIGURE 3.1 Security management cycle.

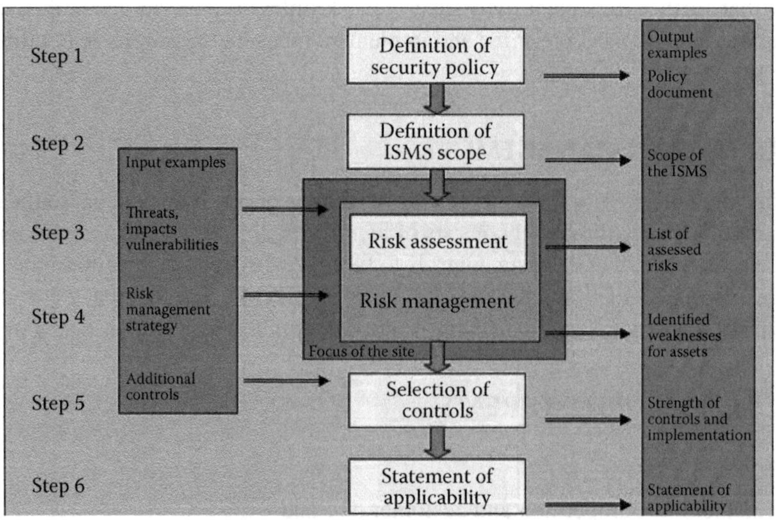

FIGURE 3.2 Risk process chart.

Hazard: safety and security; employees and equipment
Compliance: actual or potential changes in the organization's systems, processes, suppliers, etc. may create exposure to a legal or regulatory noncompliance.

RISK OPTIONS

Risk Avoidance

This is the first choice to be considered. The possibility of eliminating the existence of criminal opportunity or avoiding the creation of such an opportunity is always the best solution, when additional considerations or factors are not created as a result of this action that would create a greater risk. As an example, removing all the cash from a retail outlet would eliminate the opportunity for stealing the cash—but it would also eliminate the ability to conduct business.

Risk Reduction

When avoiding or eliminating the criminal opportunity conflicts with the ability to conduct business, the next step is the reduction of the opportunity and potential loss to the lowest level consistent with the function of the business. In the example above, the application of risk reduction might result in the business keeping only enough cash on hand for one day's operation.

Risk Spreading

Assets that remain exposed after the application of reduction and avoidance are the subjects of risk spreading. This is the concept that limits loss or potential losses by exposing the perpetrator to the probability of detection and apprehension prior to the consummation of the crime through the application of perimeter lighting, barred windows, and intrusion detection systems. The idea here is to reduce the time available to steal assets and escape without apprehension.

Risk Transfer

Risk transfer involves transferring risks to other alternatives when those risks have not been reduced to acceptable levels. The two primary methods of accomplishing risk transfer are to insure the assets or raise prices to cover the loss in the event of a criminal act. Generally speaking, when the first three steps have been properly applied, the cost of transferring risks is much lower.

Risk Acceptance

All remaining risks must simply be assumed by the business as a risk of doing business. Included with these accepted losses are deductibles that have been made as part of the insurance coverage.

SECURITY POLICY IMPLEMENTATIONS

Intrusion Detection

- Alarm devices
- Access control
- Locks, simple or sophisticated, such as biometric authentication and key-card locks
- Physical security

- Environmental elements (e.g., mountains, trees, etc.)
- Barricade
- Security guards (armed or unarmed) with wireless communication devices (e.g., two-way radio)
- Security lighting (spotlight, etc.)
- Security cameras
- Motion detectors

Procedures

- Coordination with law enforcement agencies
- Fraud management
- Risk management
- Crime Prevention Through Environmental Design (CPTED)
- Risk Analysis
- Risk mitigation
- Contingency planning

WHAT IS SECURITY?

Security is the degree of resistance to, or protection from, harm. It applies to any vulnerable and valuable asset, such as a person, dwelling, community, item, nation, or organization.

As noted by the Institute for Security and Open Methodologies in the OSSTMM 3, security provides "a form of protection where a separation is created between the assets and the threat" (*Linux Magazine*, www.linux-magazine.com). These separations are generically called "controls" and sometimes include changes to the asset or the threat.

Security is said to have two dialogues. Negative dialogue is about danger, risk, threat, etc. Positive dialogue is about opportunities, interests, profits, etc. Negative dialogue needs military equipment, armies, or police. Positive dialogue needs social capital, education, or social interaction.

Perceived Security Compared to Real Security

Perception of security may be poorly mapped to measurable objective security. For example, the fear of earthquakes has been reported to be more common than the fear of slipping on the bathroom floor, although the latter kills many more people than the former does. Similarly, the perceived effectiveness of security measures is sometimes different from the actual security provided by those measures. The presence of security protections may even be taken for the safety itself. For example, two computer security programs could be interfering with each other and even canceling each other's effect, while the owner believes he/she is getting double the protection.

Security Theater is a critical term for deployment of measures primarily aimed at raising subjective security without a genuine or commensurate concern for the effects of that action on real safety. For example, some consider the screening of airline passengers based on static databases to have been Security Theater and the

Computer Assisted Passenger Prescreening System to have created a decrease in objective security (What Is "Security Theater"? Privacy SOS, https://privacysos.org/security_theater).

Perception of security can increase objective security when it affects or deters malicious behavior, as with visual signs of security protections, such as video surveillance, alarm systems in a home, or an antitheft system in a car such as a vehicle tracking system or warning sign. Since some intruders will decide not to attempt to break into such areas or vehicles, and there can be less damage to windows in addition to protection of valuable objects inside. Without such advertisement, an intruder might, for example, approach a car, break the window, and then flee in response to an alarm being triggered. Either way, perhaps the car itself and the objects inside aren't stolen, but with perceived security, even the windows of the car have a lower chance of being damaged.

Categorizing Security

There is an immense literature on the analysis and categorization of security. Part of the reason for this is that, in most security systems, the "weakest link in the chain" is the most important. The situation is asymmetric since the "defender" must cover all points of attack while the attacker need only identify a single weak point upon which to concentrate.

Describing Categorizing Security: In preparation for selecting and specifying the appropriate security controls for organizational information systems and their respective environments of operation, organizations categorize their information and information system. To categorize the information and information system, complete the following activities:

- Information technology (IT) realm
- Computer security
- Internet security
- Application security
- Data security
- Information security
- Network security
- Endpoint security
- Physical realm
- Airport security
- Aviation security
- Communications security
- Corporate security
- Food security
- Home security
- Infrastructure security
- Physical security
- Port security/supply chain security
- Private security
- School security

- Shopping center security
- Transportation security
- Political
- National security
- Public security
- Homeland security
- Internal security
- State security
- International security
- Human security
- Monetary
- Economic security
- Financial security
- Social security

Operations security is a complement to other "traditional" security measures that evaluates the organization from an adversarial perspective.

Security Concepts

Certain concepts recur throughout different fields of security:

- *Assurance*—assurance is the level of guarantee that a security system will behave as expected.
- *Countermeasure*—a countermeasure is a way to stop a threat from triggering a risk event.
- *Defense in depth*—never rely on one single security measure alone.
- *Risk*—a risk is a possible event that could cause a loss.
- *Threat*—a threat is a method of triggering a risk event that is dangerous.
- *Vulnerability*—this is a weakness in a target that can potentially be exploited by a security threat.
- *Exploit*—a vulnerability that has been triggered by a threat—a risk of 1.0 (100%).

Home Security

Security is something applicable to all of us and involves the hardware in place on a property and personal security practices. The hardware would be the doors, locks, alarm systems, and lighting that are installed on your property. Personal security practices would be ensuring doors are locked, alarms are activated, windows are closed, and many other routine tasks that act to prevent a burglary.

Computer Security

Computer security, also known as cybersecurity or IT security, is security applied to computing devices such as computers and smartphones, as well as computer networks such as private and public networks, including the whole Internet. The field includes all five components: hardware, software, data, people, and procedures, by which digital equipment, information, and services are protected from unintended or unauthorized access, change, or destruction and is of growing importance due to the increasing reliance of computer systems in most societies. It includes physical security to prevent theft of equipment and information security to protect the data on that equipment.

Those terms generally do not refer to physical security, but a common belief among computer security experts is that a physical security breach is one of the worst kinds of security breaches as it generally allows full access to both data and equipment.

Security Management in Organizations

In the corporate world, various aspects of security are historically addressed separately—notably by distinct and often no communicating departments for IT security, physical security, and fraud prevention. Today, there is a greater recognition of the interconnected nature of security requirements, an approach variously known as holistic security, "all-hazards" management, and other terms.

3D Security

3D security is a framework promoting development, diplomacy, and defense as security strategies.

The 3D security framework recognizes that security challenges like terrorism, nuclear proliferation, global warming, and SARS or Avian Flu epidemics require a variety of tools in addressing complex threats. These tools can be categorized broadly under the headings of Development, Diplomacy, and Defense—the 3Ds of security.

3D security or "whole of government" approaches have been promoted by countries like Canada and the United Kingdom for a number of years. Now, bipartisan Congressional leaders and the Bush administration promote 3D security as a new vision for rethinking security as detailed in the 2006 National Security Strategy.

Development refers to governmental and nongovernmental efforts to build the economic, social, and political foundations of stable communities and societies. Diplomacy refers to communication or negotiation between people to solve shared problems and address conflicts through political and legal channels. Official State Department negotiations (Track I) and unofficial diplomacy (Track II) between religious, business, academic, or other civil society leaders work best on parallel tracks resulting in agreements that are legitimate, widely supported, and sustainable. Defense refers to a wide range of military tasks, including waging war, peacekeeping, or coordinating disaster response.

Security and Classified Information

Classified information is material that a government body claims is sensitive information that requires protection of confidentiality, integrity, or availability. Access is restricted by law or regulation to particular groups of people, and mishandling can incur criminal penalties and loss of respect. A formal security clearance is often required to handle classified documents or access classified data. The clearance process usually requires a satisfactory background investigation. Documents and other information assets are typically marked with one of several (hierarchical) levels of sensitivity—for example, restricted, confidential, secret, and top secret. The choice of level is often based on an impact assessment; governments often have their own set of rules, which include the levels, rules on determining the level for an information asset, and rules on how to protect information classified at each level. This often includes security clearances for personnel handling the information. Although "classified information" refers to the formal categorization and marking of material by level of sensitivity, it has also developed a sense synonymous with "censored" in U.S. English. A distinction is often made between formal security classification and privacy markings such as "commercial in confidence." Classifications can be used with additional keywords that give more detailed instructions on how data should be used or protected.

Some corporations and nongovernment organizations also assign sensitive information to multiple levels of protection, either from a desire to protect trade secrets or because of laws and regulations governing various matters such as personal privacy, sealed legal proceedings, and the timing of financial information releases.

With the passage of time, much classified information becomes much less sensitive and may be declassified and made public. Since the late 20th century, there has been freedom of information legislation in some countries, whereby the public is deemed to have the right to all information that is not considered to be damaging if released. Sometimes, documents are released with information still considered confidential obscured (redacted).

GOVERNMENT CLASSIFICATION

The purpose of classification is to protect information. Higher classifications protect information that might endanger national security. Classification formalizes what constitutes a "state secret" and accords different levels of protection based on the expected damage that the information might cause in the wrong hands.

However, classified information is frequently "leaked" to reporters by officials for political purposes. Several U.S. presidents have leaked sensitive information to get their point across to the public ("A History of How Israel Out-Foxed US Presidents," https://israelpalestinenews.org/history-israel-foxed-us-presidents).

Typical Classification Levels

Although the classification systems vary from country to country, most have levels corresponding to the following British definitions (from the highest level to lowest).

Top Secret

Top secret is the highest level of classified information. Information is further compartmented so that specific access using a code word after top secret is a legal way to hide collective and important information. Such material would cause "exceptionally grave damage" to national security if made publicly available. Prior to 1942, the United Kingdom and other members of the British Empire used Most Secret, but this was changed to match the U.S.'s top secret to simplify allied interoperability.

Secret

Secret material would cause "serious damage" to national security if it were publicly available. In the United States, operational "secret" information can be marked with an additional "LIMDIS," to limit readership.

Confidential

Confidential material would cause damage or be prejudicial to national security if publicly available.

Restricted

Restricted material would cause "undesirable effects" if publicly available. Some countries do not have such a classification; in public sectors, such as commercial industries, such a level is also called and known as "Private Information."

Official

Official material forms the generality of government business, public service delivery, and commercial activity. This includes a diverse range of information, of varying sensitivities, and with differing consequences resulting from compromise or loss. Official information must be secured against a threat model that is broadly similar to that faced by a large private company.

Unclassified

Unclassified is technically not a classification level, but this is a feature of some classification schemes, used for government documents that do not merit a particular classification or which have been declassified. This is because the information is low impact and therefore does not require any special protection, such as vetting of personnel. A plethora of pseudo-classifications exist under this category.

Clearance

Clearance is a general classification that comprises a variety of rules controlling the level of permission required to view such classified information, and how it must be stored, transmitted, and destroyed. Additionally, access is restricted on a "need-to-know" basis. Simply possessing a clearance does not automatically authorize the individual to view all material classified at that level or below that level. The individual must present a legitimate need to know in addition to the proper level of clearance.

COMPARTMENTED INFORMATION

In addition to the general risk-based classification levels, additional compartmented constraints on access exist, such as (in the U.S.) Special Intelligence (SI), which protects intelligence sources and methods; No Foreign dissemination (NOFORN), which restricts dissemination to U.S. nationals; and Originator Controlled dissemination (ORCON), which ensures that the originator can track possessors of the information. Information in these compartments is usually marked with specific keywords in addition to the classification level.

Government information about nuclear weapons often has an additional marking to show it contains such information (CNWDI).

International

When a government agency or group shares information between an agency and group of other country's government, they will generally employ a special classification scheme that both parties have previously agreed to honor.

For example, the marking ATOMAL is applied to U.S. RESTRICTED DATA or FORMERLY RESTRICTED DATA and United Kingdom ATOMIC information that has been released to NATO. ATOMAL information is marked COSMIC TOP SECRET ATOMAL (CTSA), NATO SECRET ATOMAL (NSAT), or NATO CONFIDENTIAL ATOMAL (NCA).

NATO Classifications

For example, sensitive information shared amongst NATO allies has four levels of security classification; from most to least classified:

1. COSMIC TOP SECRET (CTS),
2. NATO SECRET (NS),
3. NATO CONFIDENTIAL (NC), and
4. NATO RESTRICTED (NR).

A special case exists with regard to NATO UNCLASSIFIED (NU) information. Documents with this marking are NATO property (copyright) and must not be made public without NATO permission. In general, documents with this classification aren't cleared for Internet transmission either, unless clearly marked with RELEASABLE FOR INTERNET TRANSMISSION. Documents that can be made public, however, should be clearly marked with NON SENSITIVE INFORMATION RELEASABLE TO THE PUBLIC.

United States

Classified Information in the United States

The U.S. classification system is currently established under Executive Order 13526 and has three levels of classification—confidential, secret, and top secret. The United States had a restricted level during World War II but no longer does. U.S. regulations state that information received from other countries at the restricted level should be handled as confidential.

A variety of markings are used for material that is not classified but whose distribution is limited administratively or by other laws, for example, For Official Use Only (FOUO) or Sensitive but Unclassified (SBU). The Atomic Energy Act of 1954 provides for the protection of information related to the design of nuclear weapons. The term "Restricted Data" is used to denote certain nuclear technology.

Information about the storage, use, or handling of nuclear material or weapons is marked "Formerly Restricted Data." These designations are used in addition to level markings (Confidential, Secret and Top Secret). Information protected by the Atomic Energy Act is protected by law and information classified under the Executive Order is protected by Executive privilege.

The U.S. government insists that it is "not appropriate" for a court to question whether any document is legally classified. In the 1973 trial of Daniel Ellsberg for releasing the Pentagon Papers, the judge did not allow any testimony from Ellsberg, claiming it was "irrelevant" because the assigned classification could not be challenged. The charges against Ellsberg were ultimately dismissed after it was revealed that the government had broken the law in secretly breaking into the office of Ellsberg's psychiatrist and in tapping his telephone without a warrant. Ellsberg insists that the legal situation in the United States today is worse than it was in 1973, and Edward Snowden could not get a fair trial. The State Secrets Protection Act of 2008 might have given judges the authority to review such questions in camera, but the bill was not passed ("Trial of Daniel Ellsberg: Indictment," law2.umkc.edu/faculty/projects/FTrials/ellsberg/indictment.html).

Corporate Classification

Private corporations often require written confidentiality agreements and conduct background checks on candidates for sensitive positions. The U.S. the Employee Polygraph Protection Act prohibits private employers from requiring lie detector tests, but there are a few exceptions. Policies dictating methods for marking and safeguarding company-sensitive information (e.g., "IBM Confidential") are common, and some companies have more than one level. Such information is protected under trade secret laws. New product development teams are often sequestered and forbidden to share information about their efforts with uncleared fellow employees, the original Apple Macintosh project being a famous example. Other activities, such as mergers and financial report preparation, generally involve similar restrictions. However, corporate security generally lacks the elaborate hierarchical clearance and sensitivity structures and the harsh criminal sanctions that give government classification systems their particular tone.

Traffic Light Protocol

The Traffic Light Protocol was developed by the G8 countries to enable the sharing of sensitive information between government agencies and corporations. This protocol has now been accepted as a model for trusted information exchange by over 30 other countries. The protocol provides for four "information sharing levels" for the handling of sensitive information.

RISK

Risk is the potential of gaining or losing something of value. Values (such as physical health, social status, emotional well-being, or financial wealth) can be gained or lost when taking risk resulting from a given action or inaction, foreseen or unforeseen. Risk can also be defined as the intentional interaction with uncertainty. Uncertainty is a potential, unpredictable, and uncontrollable outcome; risk is a consequence of action taken in spite of uncertainty.

Risk perception is the subjective judgment that people make about the severity and probability of a risk and may vary person to person. Any human endeavor carries some risk, but some are much riskier than others.

Definitions

The *Oxford English Dictionary* cites the earliest use of the word in English (in the spelling of risqué from its from French original, *risqué*) as of 1621 and the spelling as risk from 1655 (https://en.oxforddictionaries.com/definition/risk). It defines risk as:

> (Exposure to) the possibility of loss, injury, or other adverse or unwelcome circumstance; a chance or situation involving such a possibility.

1. Risk is an uncertain event or condition that, if it occurs, has an effect on at least one [project] objective. (This definition, using project terminology, is easily made universal by removing references to projects.)
2. The probability of something happening multiplied by the resulting cost or benefit if it does. (This concept is more properly known as the "expectation value" or "risk factor" and is used to compare levels of risk.)
3. The probability or threat of quantifiable damage, injury, liability, loss, or any other negative occurrence that is caused by external or internal vulnerabilities, and that may be avoided through preemptive action.
4. Finance: The possibility that an actual return on an investment will be lower than the expected return.
5. Insurance: A situation where the probability of a variable (such as burning down of a building) is known but when a mode of occurrence or the actual value of the occurrence (whether the fire will occur at a particular property) is not. A risk is not an uncertainty (where neither the probability nor the mode of occurrence is known), a peril (cause of loss), or a hazard (something that makes the occurrence of a peril more likely or more severe).
6. Securities trading: The probability of a loss or drop in value. Trading risk is divided into two general categories:
    a. Systematic risk affects all securities in the same class and is linked to the overall capital-market system and therefore cannot be eliminated by diversification. Also called market risk.
    b. Nonsystematic risk is any risk that isn't market-related. Also called nonmarket risk, extra-market risk, or diversifiable risk.
7. Workplace: Product of the consequence and probability of a hazardous event or phenomenon. For example, the risk of developing cancer is estimated as the incremental probability of developing cancer over a lifetime as a result of exposure to potential carcinogens (cancer-causing substances).

Areas

Risk Assessment and Analysis

Risk Assessment and Operational Risk Management

Since risk assessment and management are essential in security management, both are tightly related. Security assessment methodologies like CCTA Risk Analysis and Management Method contain risk assessment modules as an important part of the first steps of the methodology. On the other hand, risk assessment methodologies like Mehari evolved to become security assessment methodologies. An ISO standard on risk management (Principles and guidelines on implementation) was published under code ISO 31000 on November 13, 2009.

Quantitative Analysis

There are many formal methods used to "measure" risk.

Often, the probability of a negative event is estimated by using the frequency of past similar events. Probabilities for rare failures may be difficult to estimate. This makes risk assessment difficult in hazardous industries, for example, nuclear energy, where the frequency of failures is rare, while harmful consequences of failure are severe.

Risk is often measured as the expected value of an undesirable outcome. This combines the probabilities of various possible events and some assessment of the corresponding harm into a single value. See also expected utility. The simplest case is a binary possibility of accident or no accident. The associated formula for calculating risk is then as follows:

$$R = (\text{probability of the accident occurring}) \times (\text{expected loss in case of the accident}).$$

For example, if performing activity X has a probability of 0.01 of suffering an accident of A, with a loss of 1,000, then total risk is a loss of 10, the product of 0.01 and 1,000.

Situations are sometimes more complex than the simple binary possibility case. In a situation with several possible accidents, total risk is the sum of the risks for each different accident, provided that the outcomes are comparable:

$$R = \sum \text{for all accidents (probability of the accident occurring)} \\ \times (\text{expected loss in case of the accident}).$$

SECURITY INCREASE

A security increase often occurs when a nation, state, or institution has recently suffered from a serious incident or is under the perception that there is an increased risk for an incident to occur that endangers or potentially endangers its well-being.

Historically, security has been increased by governments and private institutions for a variety of reasons, including incidents that have occurred to themselves, to other similar institutions, or to the world, nation, or region following a high-profile event or a perceived threat of one. When this occurs, many identify the move as elevated, heightened, or "beefed-up" security.

Those who are forced to make changes or be extra vigilant as a result or the increase sometimes refer to the new era as the "security age."

Institutions that often increase security in response to perceived risks include airports, government buildings, international borders, hospital, schools, religious institutions, and tourist attractions. The September 11 attacks on the World Trade Center in 2001 resulted in security being greatly increased around the world.

Common methods employed to increase security include the following:

- Increased presence of police officers and/or security guards
- Photo ID checks
- Use of surveillance (human or video)
- Use of certain instruments, like metal detectors or X-ray scanners
- Banning items seen as a potential hazard in a secured area
- Random searches
- Increased enforcement of existing rules and regulations that were previously overlooked
- Background checks
- Warning the public of the perceived threat

Surveillance

A "Nest" of Surveillance Cameras

Surveillance is the monitoring of the behavior, activities, or other changing information, usually of people for the purpose of influencing, managing, directing, or protecting them. This can include observation from a distance by means of electronic equipment (such as CCTV cameras) or interception of electronically transmitted information (such as Internet traffic or phone calls), and it can include simple, relatively no- or low-technology methods such as human intelligence agents and postal interception. Surveillance is the close observation of a person or a group by law enforcement that conducts investigations into criminal activities; continuous observation of a place, person, group, or ongoing activity in order to gather information; and use of a variety of means to collect information; an example is electronic surveillance.

Surveillance is used by governments for intelligence gathering; the prevention of crime; the protection of a process, person, group or object; or for the investigation of crime. It is also used by criminal organizations to plan and commit crimes such as robbery and kidnapping, by businesses to gather intelligence, and by private investigators.

Surveillance is often a violation of privacy and is opposed by various civil liberties groups and activists. Liberal democracies have laws that restrict domestic government and private use of surveillance, usually limiting it to circumstances where public safety is at risk. Authoritarian government seldom have any domestic restrictions; and international espionage is common among all types of countries.

Supporters of surveillance systems believe that these tools can help protect society from terrorists and criminals. They argue that surveillance can reduce crime by three means: by deterrence, by observation, and by reconstruction.

Surveillance can deter by increasing the chance of being caught and by revealing the modus operandi. This requires a minimal level of invasiveness.

Another method on how surveillance can be used to fight criminal activity is by linking the information stream obtained from them to a recognition system (for instance, a camera system that has its feed run through a facial recognition system). This can, for instance, auto-recognize fugitives and direct police to their location.

A distinction here has to be made, however, on the type of surveillance employed. Some people that say support video surveillance in city streets may not support indiscriminate telephone taps and vice versa. Besides the types, the way this surveillance is done also matters a lot; that is, indiscriminate telephone taps are supported by far fewer people than, say, telephone taps done only to people suspected of engaging in illegal activities (Electronic Surveillance, Revolvy, https://www.revolvy.com/topic/Electronic surveillance&item_type=topic).

Surveillance can also be used to give human operatives a tactical advantage through improved situational awareness or through the use of automated processes, that is, video analytics. Surveillance can help reconstruct an incident and provide evidence through the availability of footage for forensics experts. Surveillance can also influence subjective security if surveillance resources are visible or if the consequences of surveillance can be felt.

Some of the surveillance systems (such as the camera system that has its feed run through a facial recognition system mentioned above) can also have other uses besides countering criminal activity. For instance, it can help with retrieving runaway children, abducted or missing adults, and mentally disabled people.

With the advent of programs such as the Total Information Awareness program (SourceWatch, www.sourcewatch.org/index.php/Total_Information_Awareness) and ADVISE ("US Plans Massive Data Sweep," www.csmonitor.com), technologies such as high-speed surveillance computers and biometrics software, and laws such as the Communications Assistance for Law Enforcement Act (www.fcc.gov), governments now possess an unprecedented ability to monitor the activities of their subjects.

The biggest concern of detriment is securing information that is collected of those who are under total surveillance and educating the public as to why those under surveillance are being watched for the intended purpose of identifying terrorists. This is in opposition to those who use the same surveillance systems and mechanisms against civilians and have the intent to remove the laws of the land.

In addition, a significant risk of private data collection stems from the fact that this risk is too much unknown to be readily assessed today. Storage is cheap enough to have data stored forever, and the models using which it will be analyzed in a decade from now cannot reasonably be foreseen.

Countersurveillance, Inverse Surveillance, Sousveillance

Countersurveillance is the practice of avoiding surveillance or making surveillance difficult. Developments in the late 20th century have caused countersurveillance to dramatically grow in both scope and complexity, such as the Internet, increasing prevalence of electronic security systems, high-altitude (and possibly armed) unmanned aerial vehicles, and large corporate and government computer databases.

Inverse surveillance is the practice of the reversal of surveillance on other individuals or groups (e.g., citizens photographing police). Well-known examples are George Holliday's recording of the Rodney King beating and the organization Cop watch, which attempts to monitor police officers to prevent police brutality ("20 Years After Rodney King, Who's Holding Cops Accountable," www.colorlines.com/articles/20-years-after-rodney-king-whos). Countersurveillance can be used in applications to prevent corporate spying or to track other criminals by certain criminal entities. It can also be used to deter stalking methods used by various entities and organizations.

Sousveillance is inverse surveillance involving the recording by private individuals rather than government or corporate entities. Sousveillance also describes the present state of modern technological societies where anybody may take photos or videos of any person or event and then diffuse the information freely all over the world.

4 Crisis Management

Crisis management is the process by which an organization deals with a major event that threatens to harm the organization, its stakeholders, or the general public. The study of crisis management originated with the large-scale industrial and environmental disasters in the 1980s. It is considered to be the most important process in public relations.

Three elements are common to a crisis:

1. A threat to the organization
2. The element of surprise
3. A short decision time

"Crisis is a process of transformation where the old system can no longer be maintained." Therefore, the fourth defining quality is the need for change. If change is not needed, the event could more accurately be described as a failure or incident.

In contrast to risk management, which involves assessing potential threats and finding the best ways to avoid those threats, crisis management involves dealing with threats before, during, and after they have occurred. It is a discipline within the broader context of management consisting of skills and techniques required to identify, assess, understand, and cope with a serious situation, especially from the moment it first occurs to the point that recovery procedures start.

INTRODUCTION

Crisis management is a situation-based management system that includes clear roles and responsibilities and process-related organizational requirements company-wide. The response shall include action in the following areas: crisis prevention, crisis assessment, crisis handling, and crisis termination. The aim of crisis management is to be well prepared for crisis, ensure a rapid and adequate response to the crisis, maintaining clear lines of reporting and communication in the event of crisis and agreeing rules for crisis termination.

The techniques of crisis management include a number of consequent steps from the understanding of the influence of the crisis on the corporation to preventing, alleviating, and overcoming the different types of crisis.

Crisis management consists of different aspects, including the following:

- Methods used to respond to both the reality and perception of crisis.
- Establishing metrics to define what scenarios constitute a crisis and should consequently trigger the necessary response mechanisms.
- Communication that occurs within the response phase of emergency-management scenarios.

Crisis management methods provide a useful foundation for understanding terminology and frameworks relating to crisis; in this document, the focus is on the corporate exposure to risks, in particular to the black swan events that result in significant strategic threats to organizations. Crisis management is occasionally referred to as incident management.

A crisis mindset requires the ability to think of the worst-case scenario while simultaneously suggesting numerous solutions. Trial and error is an accepted discipline, as the first line of defense might not work. It is necessary to maintain a list of contingency plans and to be always on alert. Organizations and individuals should always be prepared with a rapid response plan to emergencies, which would require analysis, drills, and exercises.

The credibility and reputation of organizations is heavily influenced by the perception of their responses during crisis situations. The organization and communication involved in responding to a crisis in a timely fashion make for a challenge in businesses. There must be open and consistent communication throughout the hierarchy to contribute to a successful crisis-communication process.

The related terms "emergency management" and "business continuity management" focus, respectively, on the prompt but short-lived "first aid" type of response (e.g., putting, the fire out) and the longer-term recovery and restoration phases (e.g., moving operations to another site). Crisis is also a facet of risk management, although it is probably untrue to say that crisis management represents a failure of risk management, since it will never be possible to totally mitigate the chances of catastrophes occurring.

TYPES OF CRISIS

During the crisis management process, it is important to identify the types of crisis in that different crises necessitate the use of different crisis management strategies. Potential crises are enormous, but crises can be clustered.

The categorized eight types of crises are as follows:

1. Natural disaster
2. Technological crises
3. Confrontation
4. Malevolence
5. Organizational misdeeds
6. Workplace violence
7. Rumors
8. Terrorist attacks/manmade disasters

Natural Disaster

Natural-disaster-related crises, typically natural disasters, are such environmental phenomena as earthquakes, volcanic eruptions, tornadoes and hurricanes, floods, landslides, tsunamis, storms, and droughts that threaten life, property, and the environment itself.

Technological Crisis

Technological crises are caused by human application of science and technology. Technological accidents inevitably occur when technology becomes complex and coupled and something goes wrong in the system as a whole (technological breakdowns). Some technological crises occur when human error causes disruptions (human breakdowns). People tend to assign blame for a technological disaster because technology is subject to human manipulation, whereas they do not hold anyone responsible for natural disaster. When an accident creates significant environmental damage, the crisis is categorized as mega damage. Samples include software failures, industrial accidents, and oil spills.

Confrontation Crisis

Confrontation crises occur when discontented individuals and/or groups fight businesses, government, and various interest groups to win acceptance of their demands and expectations. The common type of confrontation crisis is boycotts, and other types are picketing, sit-ins, ultimatums to those in authority, blockade or occupation of buildings, and resisting or disobeying police.

Crisis of Malevolence

An organization faces a crisis of malevolence when opponents or miscreant individuals use criminal means or other extreme tactics for the purpose of expressing hostility or anger toward, or seeking gain from, a company, country, or economic system, perhaps with the aim of destabilizing or destroying it. Examples include product tampering, kidnapping, malicious rumors, terrorism, and espionage.

Crisis of Organizational Misdeeds

Crises occur when management takes actions it knows will harm or place stakeholders at risk for harm without adequate precautions. Three different types of crises of organizational misdeeds are crises of skewed management values, crises of deception, and crises of management misconduct.

Crisis of Skewed Management Values

Crises of skewed management values are caused when managers favor short-term economic gain and neglect broader social values and stakeholders other than investors. This state of lopsided values is rooted in the classical business creed that focuses on the interests of stockholders and tends to disregard the interests of its other stakeholders such as customers, employees, and the community.

It has three stages:

1. Precrisis
2. Acute
3. Chronic and conflict resolution

Crisis of Deception

Crises of deception occur when management conceals or misrepresents information about itself and its products in its dealing with consumers and others.

Crisis of Management Misconduct

Some crises are caused not only by skewed values and deception but also deliberate amorality and illegality.

Workplace Violence

Crises occur when an employee or former employee commits violence against other employees on organizational grounds.

Rumors

False information about an organization or its products creates crises hurting the organization's reputation. Sample is linking the organization to radical groups or stories that their products are contaminated.

Crisis Leadership

Alan Hilburg, a pioneer in crisis management, defines organizational crises as categorized as either acute crises or chronic crises (Crisis Management Explained by Alan Hilburg, http://everything.explained.today/Crisis_management/). Erika Hayes James, an organizational psychologist at the University of Virginia's Darden Graduate School of Business, identifies two primary types of organizational crisis. James (2008) defines organizational crisis as "any emotionally charged situation that, once it becomes public, invites negative stakeholder reaction and thereby has the potential to threaten the financial well-being, reputation, or survival of the firm or some portion thereof."

1. Sudden crisis
2. Smoldering crises

Sudden Crisis

Sudden crises are circumstances that occur without warning and beyond an institution's control. Consequently, sudden crises are most often situations for which the institution and its leadership are not blamed.

Smoldering Crisis

Smoldering crises differ from sudden crises in that they begin as minor internal issues that, due to manager's negligence, develop to crisis status. These are situations when leaders are blamed for the crisis and its subsequent effect on the institution in question.

There are five phases of crisis that require specific crisis leadership competencies. Each phase contains an obstacle that a leader must overcome to improve the structure and operations of an organization. The leadership competencies of integrity, positive intent, capability, mutual respect, and transparency impact the trust-building process.

1. Signal detection
2. Preparation and prevention
3. Containment and damage control
4. Business recovery
5. Learning

SIGNAL DETECTION

Sense making represents an attempt to create order and make sense, retrospectively, of what occurs. Perspective taking is the ability to consider another person's or group's point of view.

- Preparation and prevention

It is during this stage that crisis handlers begin preparing for or averting the crisis that had been foreshadowed in the signal detection stage. Using an impact/probability model allows organizations to fairly accurately predict crisis scenarios. It is recognized that the greatest organizational challenge is "speaking truth to power" to predict truly worst-case scenarios ("The Origin of the Phrase 'Speaking Truth to Power,'" http://classroom.synonym.com/origin-phrase-speaking-truth-power-11676.html). Organizations such as the Red Cross's primary mission is to prepare for and prevent the escalation of crisis events. Walmart has been described as an emergency-relief standard bearer after having witnessed the incredibly speedy and well-coordinated effort to get supplies to the Gulf Coast of the United States in anticipation of Hurricane Katrina (https://corporate.walmart.com/_news_/news-archive/2005/09/04/media-information-wal-marts-response-to-hurricane-katrina).

CONTAINMENT AND DAMAGE CONTROL

Usually the most vivid stage, the goal of crisis containment and damage control is to limit the reputational, financial, safety, and other threats to firm survival. Crisis handlers work diligently during this stage to bring the crisis to an end as quickly as possible to limit the negative publicity to the organization and move into the business recovery phase.

BUSINESS RECOVERY

When crisis hits, organizations must be able to carry on with their business in the midst of the crisis while simultaneously planning for how they will recover from the damage the crisis caused. Crisis handlers not only engage in continuity planning (determining the people, financial, and technology resources needed to keep the organization running) but will also actively pursue organizational resilience.

LEARNING

In the wake of a crisis, organizational decision makers adopt a learning orientation and use prior experience to develop new routines and behaviors that ultimately change the way the organization operates. The best leaders recognize this and are purposeful and skillful in finding the learning opportunities inherent in every crisis situation.

CRISIS COMMUNICATION

This is the effort taken by an organization to communicate with the public and stakeholders when an unexpected event occurs that could have a negative impact on the organization's reputation. This can also refer to the efforts to inform employees or the public of a potential hazard that could have a catastrophic impact.

There are three essential steps that an organization can take to prepare for and withstand a communications crisis:

1. Define your philosophy
2. Assess your vulnerabilities
3. Develop a protocol

MODELS AND THEORIES ASSOCIATED WITH CRISIS MANAGEMENT

Crisis Management Strategy

Crisis management strategy (CMS) is a corporate development strategy designed primarily to prevent crisis for follow-up company advancement. Thus, CMS is synthesis of strategic management. It includes projection of the future based on ongoing monitoring of business internal and external environment, as well as selection and implementation of crisis prevention strategy and operating management. This includes current status control based on ongoing monitoring of the internal and external environment, as well as crisis-coping strategy selection and implementation.

Crisis Management Model

Successfully managing a crisis requires an understanding of how to handle a crisis—beginning with before they occur. The arc consists of crisis avoidance, crisis mitigation, and crisis recovery.

There are three phases in any crisis management, as shown in the following:

1. The diagnosis of the impending trouble or the danger signals.
2. Choosing appropriate Turnaround Strategy.
3. Implementation of the change process and its monitoring.

Crisis Management Planning

No corporation looks forward to facing a situation that causes a significant disruption to their business, especially one that stimulates extensive media coverage. Public scrutiny can result in a negative financial, political, legal and government impact. Crisis management planning deals with providing the best response to a crisis.

Contingency Planning

Preparing contingency plans in advance, as part of a crisis-management plan, is the first step to ensuring that an organization is appropriately prepared for a crisis. Crisis-management teams can rehearse a crisis plan by developing a simulated scenario to use as a drill. The plan should clearly stipulate that the only people to speak to publicly about the crisis are the designated persons, such as the company spokesperson or crisis team members. Ideally, it should be one spokesperson who can be available on call at any time. Cooperation with the media is crucial in crisis situation, and assure that all questions are answered on time and information on what was done to resolve the situation is provided.

The first hours after a crisis breaks are the most crucial, so working with speed and efficiency is important, and the plan should indicate how quickly each function should be performed. When preparing to offer a statement externally as well as internally, information should be accurate and transparent.

Providing incorrect or manipulated information has a tendency to backfire and will greatly exacerbate the situation. The contingency plan should contain information and guidance that will help decision makers to consider not only the short-term consequences but also the long-term effects of every decision.

Business Continuity Planning

When a crisis will undoubtedly cause a significant disruption to an organization, a business continuity plan can help minimize the disruption. First, one must identify the critical functions and processes that are necessary to keep the organization running. This part of the planning should be conducted in the earliest stages and is part of a business impact analysis phase that will signpost "How much does the organization stand to lose?" Each critical function and or/process must have its own contingency plan in the event that one of the functions/processes ceases or fails, then the business/organization is more resilient, which in itself provides a mechanism to lessen the possibility of having to invoke recovery plans. Testing these contingency plans by rehearsing the required actions in a simulation will allow those involved to become more acutely aware of the possibility of a crisis. As a result, and in the event of an actual crisis, the team members will act more quickly and effectively.

A note of caution when planning training scenarios: all too often, simulations can lack ingenuity and an appropriate level of realism and as a consequence potentially lose their training value. This part can be improved by employing external exercise designers who are not part of the organizational culture and are able to test an organizations response to crisis, in order to bring about a crisis of confidence for those who manage vital systems.

Following a simulation exercise, a thorough and systematic debriefing must be conducted as a key component of any crisis simulation. The purpose of this is to create a link and draw lessons from the reality of the simulated representation and the reality of the real world.

The whole process relating to business continuity planning should be periodically reviewed to identify any number of changes that may invalidate the current plan.

Structural-Functional Systems Theory

Providing information to an organization in a time of crisis is critical to effective crisis management. The structural-functional systems theory addresses the intricacies of information networks and levels of command making up organizational communication. The structural-functional theory identifies information flow in organizations as "networks" made up of "members." Information in organizations flow in patterns called networks.

Diffusion of Innovation Theory

Another theory that can be applied to the sharing of information is the Diffusion of Innovation Theory. Developed by Everett Rogers, the theory describes how innovation is disseminated and communicated through certain channels over a period of time (https://web.stanford.edu/class/symbsys205/DiffusionofInnovations.htm).

Diffusion of innovation in communication occurs when an individual communicates a new idea to one or several others. At its most elementary form, the process involves the following:

1. An innovation
2. An individual or other unit of adoption that has knowledge of or experience with using the innovation
3. Another individual or other unit that does not yet have knowledge of the innovation
4. A communication channel connecting the two units. A communication channel is the means by which messages get from one individual to another.

ROLE OF APOLOGIES IN CRISIS MANAGEMENT

There has been debate about the role of apologies in crisis management, and some argue that apology opens an organization up for possible legal consequences: "However some evidence indicates that compensation and sympathy, two less expensive strategies, are as effective as an apology in shaping people's perceptions of the organization taking responsibility for the crisis because these strategies focus on the victims' needs. The sympathy response expresses concern for victims while compensation offers victims something to offset the suffering" (Sandman, 2005).

CRISIS LEADERSHIP

The five leadership competencies that facilitate organizational restructuring during and after a crisis are the following:

1. Building an environment of trust
2. Reforming the organization's mindset
3. Identifying obvious and obscure vulnerabilities of the organization
4. Making wise and rapid decisions as well as taking courageous action
5. Learning from crisis to effect change

Crisis leadership research concludes that leadership action in crisis reflects the competency of an organization, because the test of crisis demonstrates how well the institution's leadership structure serves the organization's goals and withstands crisis. Developing effective human resources is vital when building organizational capabilities through crisis management executive leadership.

UNEQUAL HUMAN CAPITAL THEORY

Organizational crisis can result from discrimination lawsuits. Unequal human capital and social position derives from economic theories of human and social capital concluding that minority employees receive fewer organizational rewards than those with access to executive management. In a recent study of managers in a Fortune 500 company, race was found to be a predictor of promotion opportunity or lack thereof ("Models and Theories Associated with Crisis Management," http://eng-sciencee.blogspot.com/). Thus, discrimination lawsuits can invite negative stakeholder reaction, damage the company's reputation, and threaten corporate survival.

SOCIAL MEDIA AND CRISIS MANAGEMENT

Social media has accelerated the speed that information about a crisis can spread. The viral effect of social networks such as Twitter means that stakeholders can break news faster than traditional media can—making managing a crisis harder. This can be mitigated by having the right training and policy in place as well as the right social media monitoring tools to detect signs of a crisis breaking. Social media also gives crisis management teams (CMTs) access to real-time information about how a crisis is impacting stakeholder sentiment and the issues that are of most concern to them.

Organizations should have a planned approach to releasing information to the media in the event of a crisis. A media reaction plan should include a company media representative as part of the CMT. Since there is always a degree of unpredictability during a crisis, it is best that all CMT members understand how to deal with the media and be prepared to do so, should they be thrust into such a situation.

In the face of crisis, leaders must deal with the strategic challenges they face, the political risks and opportunities they encounter, the errors they make, the pitfalls they need to avoid, and the paths away from crisis they may pursue. The necessity for management is even more significant with the advent of a 24-hour news cycle and an increasingly Internet-savvy audience with ever-changing technology at its fingertips.

Public leaders have a special responsibility to help safeguard society from the adverse consequences of crisis.

Experts in crisis management note that leaders who take this responsibility seriously would have to concern themselves with all crisis phases: the incubation stage, the onset, and the aftermath. Crisis leadership then involves five critical tasks: sense making, decision making, and meaning making, terminating, and learning.

A brief description of the five facets of crisis leadership includes the following:

1. Sense making may be considered as the classical situation assessment step in decision making.
2. Decision making is both the act of coming to a decision and the implementation of that decision.
3. Meaning making refers to crisis management as political communication.
4. Terminating a crisis is possible only if the public leader correctly handles the accountability question.
5. Learning refers to the crisis and is limited to the event and experience gained from that event. A crisis often opens a window of opportunity for reform for better or for worse.

5 Consequence Management

Consequence management, by contrast, describes ways and means to alleviate the short- and long-term physical, socioeconomic, and psychological effects of a chemical or biological attack. It describes the coordination of local, regional, national, and international assets before, during, and after an attack.

Consequence management means measures taken to restore essential operations and services in a permissive environment. It includes measures to protect public health and safety, restore essential government services, and provide emergency relief to governments, businesses, and individuals affected by the consequences of terrorism. For consequence management at the national level, the primary authority rests with the states to respond and the federal government to provide assistance as necessary.

Consequence management is predominantly an emergency management function and includes measures to protect public health and safety, restore essential government services, and provide emergency relief to governments, businesses, and individuals affected by an incident (Figure 5.1).

In an actual or potential incident or incidents involving terrorism, the Federal Emergency Management Agency (FEMA) is the federal lead for consequence management, following the Federal Response Plan. The Environmental Protection Agency's (EPA's) Criminal Enforcement program is integrated closely with and is dependent upon the Agency's Presidential Decision Directive (PDD) 39 (U.S. Policy on Counterterrorism, https://fas.org/irp/offdocs/pdd39.htm) consequence management expertise and resources, which includes numerous EPA functions, such as Solid Waste and Emergency Response, Emergency and Remedial Response and Chemical Emergency Preparedness and Prevention Office, Air and Radiation, Water and Radiation and Indoor Air. All of these functions operate under the coordinated consequence management lead of the FEMA.

What else would help our nation's communities and private sector organizations to better prepare for the threats that we face on a daily basis? The Department of Homeland Security has legislated the use of "inherently safer technologies" in response to emergency situations. Technologies exist that, had they been utilized, would have made a dramatic and positive difference in the overall preparation for, and response to, 9/11 and subsequent incidents.

What technologies would qualify for funding by Homeland Security awards and make an immediate difference in a community-wide response to a natural or manmade emergency? Clearly, there are a lot of vendors making announcements in today's marketplace on technologies related to homeland security, from chem/bio decision aids for first responders to syndromic surveillance and health alert tools, to mobile command centers filled with sophisticated communications equipment. There is no question that these products add value and individually can improve a part of the overall incident response.

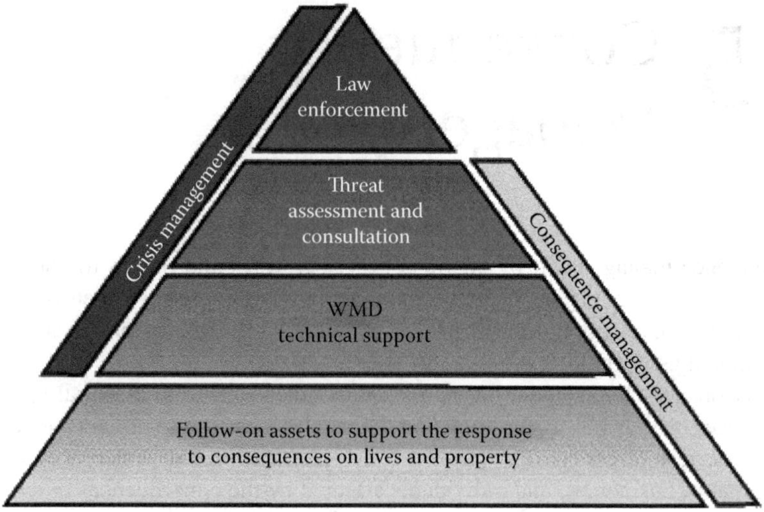

FIGURE 5.1 Crisis and consequence management.

However, it is in the area called "consequence management," where new technologies can really make a difference in the way a community responds to an emergency situation. Consequence management includes measures to protect public health and safety, restore essential services, and provide emergency relief to governments, businesses, and individuals affected by the consequences of a natural or manmade hazard. Consequence management is based upon the emergency management principles of mitigation, preparedness, response, and recovery defined by FEMA back in the 1970s with the creation of the Federal Disaster Response Plan (Resources—Federal Response Plan, www.disasters.org/emgold/frp.htm) (Figure 5.2).

True consequence management not only provides heightened preparedness for potential disasters but also focuses on improved emergency response and constant, consistent actions that mitigate the risk of emergency incidents. It also provides for coordinated efforts in recovery and remediation, from volunteer credentialing to asset management and tracking, key components in relief efforts from catastrophic emergencies, like Hurricane Katrina from last year.

FEMA further defines the components of a consequence management solution as the following.

Assessment tools:

- Hazard/Threat vulnerability and emergency readiness assessments
- Human and physical resource catalogue (emergency response/recovery assets)

Planning tools:

- Customized, hazard-specific contingency plans for emergency response
- Prebuilt standard operating procedure templates/checklists

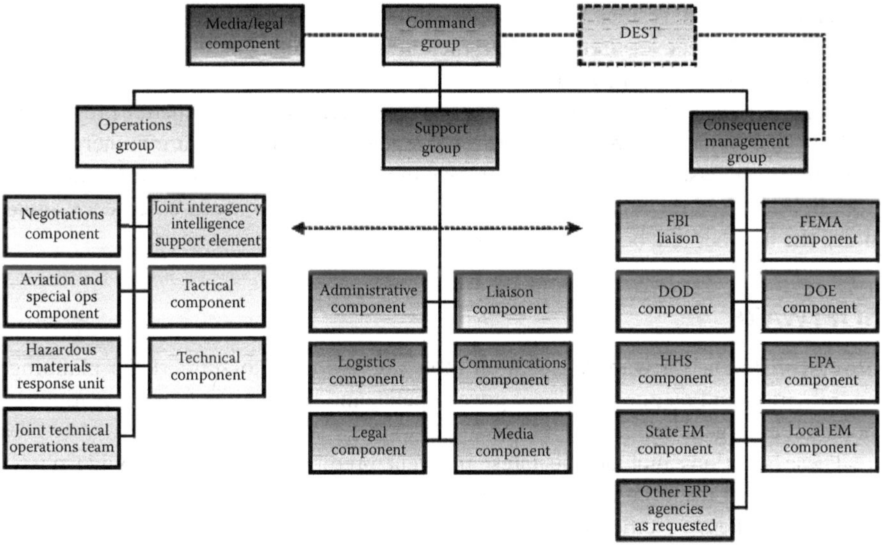

FIGURE 5.2 ICS chart. DOD, Department of Defense; DOE, Department of Energy; EM, emergency management; EPA, Environmental Protection Agency; FBI, Federal Bureau of Investigation; FEMA, Federal Emergency Management Agency; FRP, Federal Response Plan; HHS, Human & Health Services; ICS, Incident Command System.

- Geomapping and situational awareness (tools for mission planning, asset tracking, and training/exercises)
- Communication and alert tools
- Risk surveillance and event prediction
- Real-time alert dissemination with chain-of-command notification
- Medical response tools
- Syndromic surveillance tied to point of care/incident clinical documentation
- C4I Tools (Command, Control, Communications, Computers, and Intelligence)
- Command, control, communication, coordination, and information

Recovery tools:

- Damage assessment and reimbursement
- Volunteer management/credentialing and asset management (mutual aid)

The above components combine to form a solution that provides hazard vulnerability and emergency readiness assessments for any type of emergency (an all-hazard approach) followed by response plans jointly developed by government and industry (mutual aid) and improved alert dissemination and communication leading to coordinated medical response and overall incident command and control, with a focus on improving the preparedness and response efforts of first responders (firefighters, policemen, and emergency medical technicians) and local emergency managers (trained incident commanders who typically manage the emergency operations center (EOC) that resides in each community). The solution is completed with components for recovery/remediation, including the management of volunteers and assets from lending organizations.

The need for a complete consequence management solution to better prepare for and respond to natural and manmade emergencies is clear. There are numerous and meaningful, measurable benefits to improving command, control, communication, coordination, and recovery within each community for all-hazard emergency situations, the number one benefit being the saving of lives.

SOFTWARE

Several different software automation tools are available to support the planning, coordination and response of local governments and private sector organizations to potential emergencies.

These solutions include the latest in event prediction, community-wide physical and human resource identification and management, automation of contingency plans, awareness of emergency situations via geomapping, real-time interagency communication, and overall incident response task workflow and management. Consequence management technologies serve to help our communities establish control over their resources, effectively plan for emergency situations, and improve the appropriate response to an event, thereby mitigating risk and saving lives.

So what overall functionality should a true consequence management solution provide (Figure 5.3)?

1. Event prediction and alerts (real-time identification and visualization of resource status, availability and location, and rapid risk communication/alert dissemination to key personnel as part of a simulation/drill exercise, or prior to an actual event)
   - Monitoring of multiple information sources for possible alerts to response participants (via existing external system interfaces)
   - Rapid risk communication and alerts regarding an identified threat (reaching response participants via e-mail, fax, phone, Internet, cellular, wireless and handheld connectivity and connecting with existing information sources, including EOCs, fire, police, emergency medical service, e911, hospitals, public works and even federal/state agencies)
   - Preevent polling and visualization of physical and human resources (status, availability, and location)
2. Contingency planning (identification of physical and human resources by type and workflow, threat response planning, graphical design of response plans, and automation of resource and response plan templates/checklists)
   - Human and physical resource inventory automation and cataloguing
   - Geomapping and situational awareness (asset tracking, mission planning, and incident management tools)
   - Prebuilt response templates/checklists (natural and manmade hazards)
   - Response template/checklist design tool (to create custom templates)
   - Public/private sector response communication and coordination

Consequence = impact or severity of the effect		
Score	**Impact**	**Descriptor**
1	Insignificant	• Negligible effects • Strategic view: normal difficulties o Stakeholder faith affected lasting less than 6 months o Isolated injury o Financial loss of less than $5K
2	Minor	• Normal administrative difficulties • Strategic view: delay will occur in fulfilling objective o Stakeholder faith affected lasting longer than 6 months o Isolated injury o Financial less than $1M
3	Significant	• Delay in accomplishing program or project o Stakeholder faith affected lasting longer than 12 months o Multiple injury o System interruption o Dispute that could affect term o Financial loss greater than $1M less than $2M
4	Major	• Program or project redesign required, re-approval and or re-do required • Strategic view: integrated plan timeline affected o Stakeholder faith affected lasting longer than 18 months o Isolated loss of life o Major system loss at critical time o Dispute that could affect term o Financial loss greater than $2M less than $5M
5	Severe/ catastrophic	• Project or program irrevocable finished, objective not met • Strategic view: mandate or objective not met o Stake holder faith affected lasting longer than 24 months o Multiple loss of life o Complete system crash o Dispute that could cause loss of full term o Inability to recruit students or staff o Financial loss greater than $5M

FIGURE 5.3 Consequence impact/severity.

3. Consequence coordination and response (actual deployment/coordination of resources during an event, rapid risk communication/alert dissemination to response participants, activation of response plans, prompts to users for specific actions, embedded escalation of alerts, and prompts and full audit/documentation of actions taken by participants during the event)
   - Real-time identification/polling and visualization of physical or human resource status, location, and availability
   - Real-time alert facility to human participants with embedded workflow, timers, and escalation options
   - Resource deployment and optimization (automated prompts for users to take appropriate actions, and for specific allocation/coordination of resources across emergency service disciplines)
   - Rapid risk communication/alert information dissemination (to the media and the public)
   - Command center and control via a graphical user interface (geo-mapping, top-level view of the incident with drill-down to specific

components of the response—also allows user input regarding response components that may change a component's action within the response, or the overall response template)
- Real-time suggested actions for all response participants
- Full audit trail of actual actions performed by response participants
- All functionality available in a "live" incident response is available in a full response simulation
- Real-time event audit and "live" documentation of the entire incident response scenario during the simulation

4. Postevent audit/documentation (including resource optimization, recommendations for future events, actual vs. plan reports, and non-emergency coordination such as logistics)
   - Graphical reports (standard line/pie/bar graphing tool) of actual versus plan variances
   - Geomapping summary of incident response (geographic information system mapping and trend analysis, with a "playback" of asset movements and actions taken by emergency service discipline)
   - Customized regulatory reports (National Response Center, Environmental Protection Agency, Centers for Disease Control and Prevention, Department of Transportation, etc.)
   - Postevent resource deployment and coordination planning
5. Recovery/remediation initiatives (including mutual aid initiatives such as volunteer credentialing and management; asset identification, coordination, and tracking; damage assessment; and interagency postevent coordination)
   - Volunteer credentialing/management (technologies for identifying persons by role/function, credentialing them, and managing their efforts)
   - Asset management (identification of available assets for recovery efforts, obtaining, activating, and tracking these resources)
   - Damage assessment (automated documentation of post-event damage, and submission for financial reimbursement)
   - Interagency collaboration (postevent planning and coordination between agencies on recovery initiatives)
6. Simulation and drill development (automated, real-time coordination of the critical exercises that can expose gaps in response plans, improve response time, and mitigate risk to actual events)
   - A library of simulations (templates)
   - A simulation development wizard (creating custom drills for multiple threats)
   - Simulation execution and management (real-time, automated drills with full audit functions by human and physical resource)
   - Visualization of the exercise (mission planning and situational awareness of the exercise incident and assets utilized in the exercise via geomapping tools)
   - Postdrill reports

BENEFITS OF CONSEQUENCE MANAGEMENT SYSTEMS

There are numerous benefits for local communities in identifying and implementing a complete consequence management solution. Among these are the following:

- Day-to-day value in improved communication among community agencies
- Faster, accurate, more precise alerts to key personnel regarding potential emergencies
- Improved preparedness in advance of potential disasters
- Better coordination among agencies during the response to an incident
- Real-time, accurate information regarding status of human and physical resources responding to an emergency
- Optimized deployment, control, and coordination of resources in the field
- Full, real-time audit and documentation of actions taken by incident responders
- Full documentation of mutual-aid requests, and the community response to these requests
- Faster recovery from the event
- Complete control and coordination of simulation/drill exercises to identify weaknesses in emergency response plans by hazard type and to ensure full education/preparedness of response participants
- Ability to document improvements in incident response over time
- A clear indication of positive actions taken to respond to the threat of terrorism, and to protect the safety of citizens within the community
- Proof of steps taken to improve preparedness for, and response to, virtually any kind of emergency situation
- Lives saved as a result of a faster, coordinated disaster response

The greatest benefit to local communities will be in the adoption of solutions that can facilitate the overall incident response to any natural or manmade hazard. These consequence management solutions will provide numerous, quantifiable benefits for U.S. communities and their corporate citizens by mitigating the risks associated with emergency situations, better preparing the community and private sector organizations for these events, vastly improving the response to natural and manmade incidents and facilitating a smooth recovery from the event.

The need for leadership, expertise, and resources in identifying and responding to both natural and manmade threats has become more clearly understood in the aftermath of 9/11 and the recent hurricane season. The next task for every community is to define the distinct roles of first responders, the medical community, public health professionals, and all other participants in a community-based response system; get them to focus on interagency collaboration; cross-train them on emergency management principles and the Incident Command System (ICS); and to define emergency response plans based on an all-hazard approach—and to do so before the next event occurs.

Natural disasters and terrorism do not respect geographic boundaries, and we now know that each and every community in this nation is vulnerable to a natural or

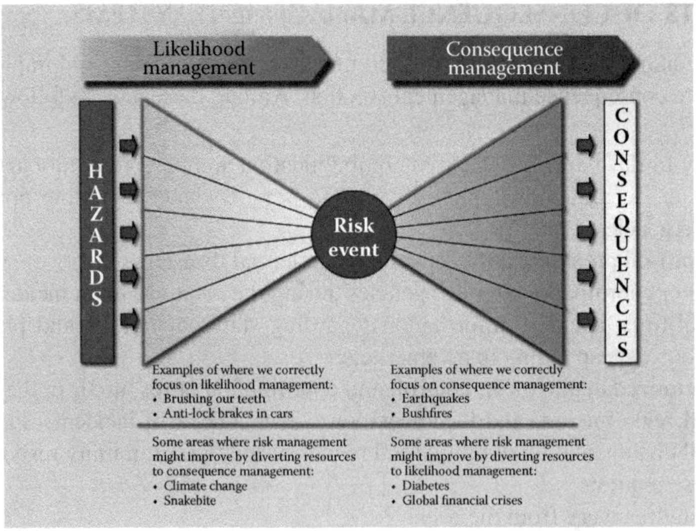

FIGURE 5.4 Risk event.

manmade disaster. We may not be able to prevent these incidents, but we can minimize the effects of these emergencies, both on an economic and human scale, with the adoption of technologies that provide consequence management solutions, and with a focus on interagency collaboration (Figure 5.4).

COORDINATING RESPONSE

Information sharing and interagency coordination are clearly needed to facilitate a successful emergency incident response. Yet many public and private organizations lack the comprehensive emergency response plans that define the roles and responsibilities of trained personnel responding to an unexpected incident and describe how to work "side by side" with responders from other agencies.

Training itself is an issue, as many organizations do not know where to turn for assistance regarding emergency preparedness, nor do they have the time to stop the daily task of operating a business or service to educate personnel on how to respond to disasters or to implement the emergency preparedness and response requirements now mandated by the federal government.

In the local communities, where such training is mandated, agencies participating in an emergency response are often not coordinated in their efforts, and, to make matters worse, severe shortages exist in the area of emergency management personnel. The preparedness and response expertise reflected in this type of human capital is in great demand post-9/11, but very hard to find.

Clearly, the response to a bioterrorism incident would be improved with better collaboration and coordination among the agencies and private companies participating in this response effort. How is this accomplished? With training on the principles of emergency management and the ICS established by FEMA for use in response to any type of hazard (an all-hazard approach) and the implementation of "inherently safer technologies for improved incident response," advocated by the Department of Homeland Security.

The all-hazard approach has been a cornerstone of FEMA's response program since the agency was first established. It integrates the various emergency plans and activities into a "life cycle" of mitigation, preparedness, response, and recovery (the principles of emergency management) and, when combined with the ICS, provides a template for interagency coordination that is directly applicable to events stemming from chemical and biological hazards, as well as all other manmade or natural events.

Every community with an EOC has "in-house" expertise on the all-hazard approach, which should be utilized to assist other agencies (such as public health) in the assessment, planning, and simulation of a community-based response to emergency situations.

Each and every year, cities across the nation demonstrate that collaboration between agencies, using the all-hazard approach to fully prepare the communities for a coordinated incident response. This example refers to communities that are within a 10-mile radius of a nuclear power facility, where radiological emergency preparedness exercises are performed.

Horizontal communication and rapid exchange of information among agencies are a basic requirement during any emergency, and the all-hazard approach has proven to be a successful response system for both natural and manmade events. Each community should leverage their experience and knowledge of the all-hazard response system and ensure that all agencies participating in an emergency incident response are cross-trained in this approach.

Summarizing about what "consequence management" is, it was introduced into the national security lexicon with the promulgation of PDD 39 in 1995. This PDD's purpose was to establish how the nation would respond to terrorism employing weapons of mass destruction (WMD) and how the consequences of such an incident would be managed. Despite the very broad outlines of the PDD, consequence management, for the most part, continues to be a tabula rasa. It has never been done before. There are not yet any experts. The concept of "consequence management" will remain ill-defined as well-intentioned organizations and individuals grapple with how they might institutionalize a comprehensive operational response to a terrorist's use of WMD. Make no mistake, though, consequence management is not only a characteristic of the post-Cold War, it is also an indicator of just how capable this nation is of reconceptualizing and reorganizing for its security in the 21st century.

6 Risk Management

Risk management is the identification, assessment, and prioritization of risks (defined in ISO 31000 as the effect of uncertainty on objectives) followed by coordinated and economical application of resources to minimize, monitor, and control the probability and/or impact of unfortunate events or to maximize the realization of opportunities (Risk Management—Principles and Guidelines, www.iso.org). Risk management's objective is to assure uncertainty does not deflect the endeavor from the business goals.

Risks can come from various sources, including uncertainty in financial markets, threats from project failures (at any phase in design, development, production, or sustainment life cycles), legal liabilities, credit risk, accidents, natural causes and disasters, deliberate attack from an adversary, or events of uncertain or unpredictable root-cause. There are two types of events; that is, negative events can be classified as risks, while positive events are classified as opportunities. Several risk management standards have been developed, including the Project Management Institute, the National Institute of Standards and Technology, actuarial societies, and International Organization for Standardization (ISO) standards. Methods, definitions, and goals vary widely according to whether the risk management method is in the context of project management, security, engineering, industrial processes, financial portfolios, actuarial assessments, or public health and safety.

Risk sources are identified and located in human factor variables, mental states, and decision making as well as infrastructural or technological assets and tangible variables.

The interaction between human factors and tangible aspects of risk highlights the need to focus closely on human factors as one of the main drivers for risk management, a "change driver" that comes first of all from the need to know how humans perform in challenging environments and in face of risks.

It is an extremely hard task to be able to apply an objective and systematic self-observation and to make a clear and decisive step from the level of the mere "sensation" that something is going wrong to the clear understanding of how, when, and where to act. The truth of a problem or risk is often obfuscated by wrong or incomplete analyses, fake targets, perceptual illusions, unclear focusing, altered mental states, and lack of good communication and confrontation of risk management solutions with reliable partners. This makes the human factor aspect of risk management sometimes heavier than its tangible and technological counterpart.

Strategies to manage threats (uncertainties with negative consequences) typically include avoiding the threat, reducing the negative effect or probability of the threat, transferring all or part of the threat to another party, and even retaining some or all of the potential or actual consequences of a particular threat, and the opposites for opportunities (uncertain future states with benefits).

In ideal risk management, a prioritization process is followed whereby the risks with the greatest loss (or impact) and the greatest probability of occurring are handled first, and risks with lower probability of occurrence and lower loss are handled in descending order.

In practice, the process of assessing overall risk can be difficult, and balancing resources used to mitigate between risks with a high probability of occurrence but lower loss versus a risk with high loss but lower probability of occurrence can often be mishandled.

Intangible risk management identifies a new type of a risk that has a 100% probability of occurring but is ignored by the organization due to a lack of identification ability. For example, when deficient knowledge is applied to a situation, a knowledge risk materializes. Relationship risk appears when ineffective collaboration occurs. Process-engagement risk may be an issue when ineffective operational procedures are applied. These risks directly reduce the productivity of knowledge workers and decrease cost-effectiveness, profitability, service, quality, reputation, brand value, and earnings quality. Intangible risk management allows risk management to create immediate value from the identification and reduction of risks that reduce productivity.

Risk management also faces difficulties in allocating resources. This is the idea of opportunity cost. Resources spent on risk management could have been spent on more profitable activities. Again, ideal risk management minimizes spending (or manpower or other resources) and also minimizes the negative effects of risks.

According to the definition of risk, the risk is the possibility that an event will occur and adversely affect the achievement of an objective. Therefore, risk itself has uncertainty. Risk management such as Committee of Sponsoring Organizations (COSO) Enterprise Risk Management (ERM)–Integrated Framework can help managers have a good control for their risk.

Each company may have different internal control components, which leads to different outcomes. For example, the framework for enterprise risk management components includes Internal Environment, Objective Setting, Event Identification, Risk Assessment, Risk Response, Control Activities, Information and Communication, and Monitoring.

METHOD

For the most part, these methods consist of the following elements, performed, more or less, in the following order.

1. Identify and characterize threats.
2. Assess the vulnerability of critical assets to specific threats.
3. Determine the risk (i.e., the expected likelihood and consequences of specific types of attacks on specific assets).
4. Identify ways to reduce those risks.
5. Prioritize risk-reduction measures based on a strategy.

PRINCIPLES OF RISK MANAGEMENT

The ISO identifies the following principles of risk management:

Risk management should:

- Create value—resources expended to mitigate risk should be less than the consequence of inaction.
- Be an integral part of organizational processes.

- Be part of decision making process.
- Explicitly address uncertainty and assumptions.
- Be a systematic and structured process.
- Be based on the best available information.
- Be tailorable.
- Take human factors into account.
- Be transparent and inclusive.
- Be dynamic, iterative, and responsive to change.
- Be capable of continual improvement and enhancement.
- Be continually or periodically reassessed.

PROCESS

According to the standard ISO 31000 "Risk management—Principles and guidelines on implementation," the process of risk management consists of several steps as follows (www.iso.org).

Establishing the Context

This involves the following:

1. Identification of risk in a selected domain of interest
2. Planning the remainder of the process
3. Mapping out the following: the social scope of risk management
4. The identity and objectives of stakeholders
5. The basis upon which risks will be evaluated, constraints
6. Defining a framework for the activity and an agenda for identification
7. Developing an analysis of risks involved in the process
8. Mitigation or solution of risks using available technological, human and organizational resources.

Identification

After establishing the context, the next step in the process of managing risk is to identify potential risks. Risks are about events that, when triggered, cause problems or benefits. Hence, risk identification can start with the source of our problems and those of our competitors (benefit), or with the problem itself.

- *Source analysis*—Risk sources may be internal or external to the system that is the target of risk management (use mitigation instead of management since by its own definition risk deals with factors of decision making that cannot be managed).
- *Problem analysis*—Risks are related to identified threats. For example: the threat of losing money, the threat of abuse of confidential information or the threat of human errors, accidents, and casualties. The threats may exist with various entities, most important with shareholders, customers, and legislative bodies such as the government.

When either source or problem is known, the events that a source may trigger or the events that can lead to a problem can be investigated. For example: stakeholders withdrawing during a project may endanger funding of the project; confidential information may be stolen by employees even within a closed network; lightning striking an aircraft during takeoff may make all people on board immediate casualties.

The chosen method of identifying risks may depend on culture, industry practice, and compliance. The identification methods are formed by templates or the development of templates for identifying source, problem, or event. Common risk identification methods are as follows:

- *Objectives-based risk identification*—Organizations and project teams have objectives. Any event that may endanger achieving an objective partly or completely is identified as risk.
- *Scenario-based risk identification*—In scenario analysis, different scenarios are created. The scenarios may be the alternative ways to achieve an objective, or an analysis of the interaction of forces in, for example, a market or battle. Any event that triggers an undesired scenario alternative is identified as risk.
- *Taxonomy-based risk identification*—The taxonomy in taxonomy-based risk identification is a breakdown of possible risk sources. Based on the taxonomy and knowledge of best practices, a questionnaire is compiled. The answers to the questions reveal risks.
- *Common-risk checking*—In several industries, lists with known risks are available. Each risk in the list can be checked for application to a particular situation.
- *Risk charting*—This method combines the above approaches by listing resources at risk, threats to those resources, and modifying factors that may increase or decrease the risk and consequences it is wished to avoid. Creating a matrix under these headings enables a variety of approaches. One can begin with resources and consider the threats they are exposed to and the consequences of each. Alternatively, one can start with the threats and examine which resources they would affect, or one can begin with the consequences and determine which combination of threats and resources would be involved to bring them about.

Assessment

Once risks have been identified, they must then be assessed as to their potential severity of impact (generally a negative impact, such as damage or loss) and to the probability of occurrence. These quantities can be either simple to measure, in the case of the value of a lost building, or impossible to know for sure in the case of an unlikely event, the probability of occurrence of which is unknown. Therefore, in the assessment process, it is critical to make the best educated decisions in order to properly prioritize the implementation of the risk management plan.

Even a short-term positive improvement can have long-term negative impacts. Take the "turnpike" example. A highway is widened to allow more traffic. More traffic capacity leads to greater development in the areas surrounding the improved traffic capacity. Over time, traffic thereby increases to fill available capacity. Turnpikes thereby need to be expanded in a seemingly endless cycles. There are many other engineering examples where expanded capacity (to do any function) is soon filled by increased demand. Since expansion comes at a cost, the resulting growth could become unsustainable without forecasting and management.

The fundamental difficulty in risk assessment is determining the rate of occurrence since statistical information is not available on all kinds of past incidents and is particularly scanty in the case of catastrophic events, simply because of their infrequency. Furthermore, evaluating the severity of the consequences (impact) is often quite difficult for intangible assets. Asset valuation is another question that needs to be addressed.

Thus, best educated opinions and available statistics are the primary sources of information. Nevertheless, risk assessment should produce such information for senior executives of the organization that the primary risks are easy to understand and that the risk management decisions may be prioritized within overall company goals. Thus, there have been several theories and attempts to quantify risks. Numerous different risk formulae exist, but perhaps the most widely accepted formula for risk quantification is the following: rate (or probability) of occurrence multiplied by the impact of the event equals risk magnitude.

Composite Risk Index

The above formula can also be rewritten in terms of a composite risk index, as follows:

Composite risk index = impact of risk event × probability of occurrence.

The impact of the risk event is commonly assessed on a scale of 1 to 5, where 1 and 5 represent the minimum and maximum possible impact, respectively, of an occurrence of a risk (usually in terms of financial losses). However, the 1 to 5 scale can be arbitrary and need not be on a linear scale.

The probability of occurrence is likewise commonly assessed on a scale from 1 to 5, where 1 represents a very low probability of the risk event actually occurring while 5 represents a very high probability of occurrence. This axis may be expressed in either mathematical terms (event occurs once a year, once in ten years, once in 100 years, etc.) or may be expressed in "plain English" (event has occurred here very often, event has been known to occur here, event has been known to occur in the industry, etc.). Again, the 1 to 5 scale can be arbitrary or nonlinear depending on decisions by subject matter experts.

The composite risk index can thus take values ranging (typically) from 1 through 25, and this range is usually arbitrarily divided into three subranges. The overall risk assessment is then low, medium, or high, depending on the subrange containing the calculated value of the Composite Index.

Likewise, the impact of the risk is not easy to estimate since it is often difficult to estimate the potential loss in the event of risk occurrence.

Further, both the above factors can change in magnitude depending on the adequacy of risk avoidance and prevention measures taken and due to changes in the external business environment. Hence, it is absolutely necessary to periodically reassess risks and intensify/relax mitigation measures, or as necessary. Changes in procedures, technology, schedules, budgets, market conditions, political environment, or other factors typically require reassessment of risks.

Risk Options

Risk mitigation measures are usually formulated according to one or more of the following major risk options:

- Design a new business process with adequate built-in risk control and containment measures from the start.
- Periodically reassess risks that are accepted in ongoing processes as a normal feature of business operations and modify mitigation measures.
- Transfer risks to an external agency (e.g., an insurance company).
- Avoid risks altogether (e.g., by closing down a particular high-risk business area).

Later research has shown that the financial benefits of risk management are less dependent on the formula used but are more dependent on the frequency and how risk assessment is performed.

Potential Risk Treatments

Once risks have been identified and assessed, all techniques to manage the risk fall into one or more of these four major categories:

1. Avoidance (eliminate, withdraw from or not become involved)
2. Reduction (optimize—mitigate)
3. Sharing (transfer—outsource or insure)
4. Retention (accept and budget)

Ideal use of these risk control strategies may not be possible. Some of them may involve trade-offs that are not acceptable to the organization or person making the risk management decisions.

Risk Avoidance

This includes not performing an activity that could carry risk. An example would be not buying a property or business in order to not take on the legal liability that comes with it. Another would be not flying in order not to take the risk that the airplane were to be hijacked. Avoidance may seem the answer to all risks, but avoiding risks also means losing out on the potential gain that accepting (retaining) the risk may have allowed. Not entering a business to avoid the risk of loss also avoids the possibility of earning profits. Increasing risk regulation in hospitals has led to avoidance of treating higher risk conditions, in favor of patients presenting with lower risk (Risk Management—Bolds Risk, https://boldrisk.com/risk-management).

Hazard Prevention

Hazard prevention refers to the prevention of risks in an emergency. The first and most effective stage of hazard prevention is the elimination of hazards. If this takes too long, is too costly, or is otherwise impractical, the second stage is mitigation.

Risk Reduction

Risk reduction or "optimization" involves reducing the severity of the loss or the likelihood of the loss from occurring. For example, sprinklers are designed to put out a fire to reduce the risk of loss by fire. Acknowledging that risks can be positive or negative, optimizing risks means finding a balance between negative risk and the benefit of the operation or activity and between risk reduction and effort applied. Outsourcing could be an example of risk reduction if the outsourcer can demonstrate higher capability at managing or reducing risks.

Risk Sharing

Briefly defined as "sharing with another party the burden of loss or the benefit of gain, from a risk, and the measures to reduce a risk" (Aujero, 2014).

The term of "risk transfer" is often used in place of risk sharing in the mistaken belief that you can transfer a risk to a third party through insurance or outsourcing. In practice, if the insurance company or contractor go bankrupt or end up in court, the original risk is likely to still revert to the first party. As such, in the terminology of practitioners and scholars alike, the purchase of an insurance contract is often described as a "transfer of risk." However, technically speaking, the buyer of the contract generally retains legal responsibility for the losses "transferred," meaning that insurance may be described more accurately as a postevent compensatory mechanism. For example, a personal injuries insurance policy does not transfer the risk of a car accident to the insurance company. The risk still lies with the policy holder, namely, the person who has been in the accident. The insurance policy simply provides that if an accident (the event) occurs involving the policy holder, then some compensation may be payable to the policy holder that is commensurate with the suffering/damage.

Some ways of managing risk fall into multiple categories. Risk retention pools are technically retaining the risk for the group, but spreading it over the whole group involves transfer among individual members of the group. This is different from traditional insurance, in that no premium is exchanged between members of the group up front, but instead losses are assessed in all members of the group.

Risk Retention

Risk retention involves accepting the loss, or benefit of gain, from a risk when it occurs. True self-insurance falls in this category. Risk retention is a viable strategy for small risks where the cost of insuring against the risk would be greater over time than the total losses sustained. All risks that are not avoided or transferred are retained by default.

This includes risks that are so large or catastrophic that they either cannot be insured against or the premiums would be infeasible. War is an example since most property and risks are not insured against war, so the loss attributed by war is retained by the insured.

Also any amounts of potential loss (risk) over the amount insured are retained risk. This may also be acceptable if the chance of a very large loss is small or if the cost to insure for greater coverage amounts is so great it would hinder the goals of the organization too much. Risk retention or acceptance is common type of risk response on treats and opportunities.

Risk Management Plan

Select appropriate controls or countermeasures to measure each risk. Risk mitigation needs to be approved by the appropriate level of management. For instance, a risk concerning the image of the organization should have top management decision behind it, whereas information technology management would have the authority to decide on computer virus risks.

The risk management plan should propose applicable and effective security controls for managing the risks. For example, an observed high risk of computer viruses could be mitigated by acquiring and implementing antivirus software. A good risk management plan should contain a schedule for control implementation and responsible persons for those actions.

According to ISO/IEC 27001 (www.iso.org), the stage immediately after completion of the risk assessment phase consists of preparing a Risk Treatment Plan, which should document the decisions about how each of the identified risks should be handled. Mitigation of risks often means selection of security controls, which should be documented in a Statement of Applicability, which identifies which particular control objectives and controls from the standard have been selected, and why.

IMPLEMENTATION

Implementation follows all of the planned methods for mitigating the effect of the risks. Purchase insurance policies for the risks that have been decided to be transferred to an insurer, avoid all risks that can be avoided without sacrificing the entity's goals, reduce others, and retain the rest.

Review and Evaluation of the Plan

Initial risk management plans will never be perfect. Practice, experience, and actual loss results will necessitate changes in the plan and contribute information to allow possible different decisions to be made in dealing with the risks being faced.

Risk analysis results and management plans should be updated periodically. There are two primary reasons for this:

1. To evaluate whether the previously selected security controls are still applicable and effective.
2. To evaluate the possible risk level changes in the business environment. For example, information risks are a good example of rapidly changing business environment.

Limitations

Prioritizing the risk management processes too highly could keep an organization from ever completing a project or even getting started. This is especially true if other work is suspended until the risk management process is considered complete.

It is also important to keep in mind the distinction between risk and uncertainty. Risk can be measured by impacts × probability.

If risks are improperly assessed and prioritized, time can be wasted in dealing with risk of losses that are not likely to occur. Spending too much time assessing and managing unlikely risks can divert resources that could be used more profitably. Unlikely events do occur, but if the risk is unlikely enough to occur, it may be better to simply retain the risk and deal with the result if the loss does in fact occur. Qualitative risk assessment is subjective and lacks consistency. The primary justification for a formal risk assessment process is legal and bureaucratic.

Hazard Analysis

A hazard analysis is used as the first step in a process used to assess risk. The result of a hazard analysis is the identification of different type of hazards. A hazard is a potential condition and exists or not (probability is 1 or 0). It may in single existence or in combination with other hazards (sometimes called events) and conditions become an actual functional failure or accident (mishap). The way this exactly happens in one particular sequence is called a scenario.

This scenario has a probability (between 1 and 0) of occurrence. Often, a system has many potential failure scenarios. It also is assigned a classification, based on the worst case severity of the end condition. Risk is the combination of probability and severity. Preliminary risk levels can be provided in the hazard analysis.

The validation, more precise prediction (verification), and acceptance of risk are determined in the risk assessment (analysis). The main goal of both is to provide the best selection of means of controlling or eliminating the risk. The term is used in several engineering specialties, including avionics, chemical process safety, safety engineering, reliability engineering, and food safety.

Hazards and Risk

A hazard is defined as a "Condition, event, or circumstance that could lead to or contribute to an unplanned or undesirable event" (Federal Aviation Administration, 2017). Seldom does a single hazard cause an accident or a functional failure. More often, an accident or operational failure occurs as the result of a sequence of causes. A hazard analysis will consider system state, for example, operating environment, as well as failures or malfunctions.

While in some cases, safety or reliability risk can be eliminated, in most cases, a certain degree of risk must be accepted. In order to quantify expected costs before the fact, the potential consequences and the probability of occurrence must be considered. Assessment of risk is made by combining the severity of consequence with the likelihood of occurrence in a matrix. Risks that fall into the "unacceptable" category (e.g., high severity and high probability) must be mitigated by some means to reduce the level of safety risk.

IEEE STD-1228-1994 Software Safety Plans (standards.ieee.org/findstds/standard/1228-1994.html) prescribes industry best practices for conducting software safety hazard analyses to help ensure that safety requirements and attributes are defined and specified for inclusion in software that commands, controls, or monitors critical functions. When software is involved in a system, the development and design assurance of that software are often governed by DO-178B.* The severity of consequence identified by the hazard analysis establishes the criticality level of the software. Software criticality levels range from A to E, corresponding to the severity of Catastrophic to No Safety Effect. Higher levels of rigor are required for level A and B software, and corresponding functional tasks and work products in the system safety domain are used as objective evidence of meeting safety criteria and requirements.

SEVERITY DEFINITIONS—SAFETY RELATED

Severity	Definition
Catastrophic	Results in multiple fatalities and/or loss of the system
Hazardous	Reduces the capability of the system or the operator ability to cope with adverse conditions to the extent that there would be • Large reduction in safety margin or functional capability • Crew physical distress/excessive workload such that operators cannot be relied upon to perform required tasks accurately or completely • Serious or fatal injury to small number of occupants of aircraft (except operators) • Fatal injury to ground personnel and/or general public

(Continued)

* DO-178B, Software Considerations in Airborne Systems and Equipment Certification, is a guideline dealing with the safety of safety-critical software used in certain airborne systems. Although technically a guideline, it is (or was) a de facto standard for developing avionics software systems.

Major	Reduces the capability of the system or the operators to cope with adverse operating conditions to the extent that there would be • Significant reduction in safety margin or functional capability • Significant increase in operator workload • Conditions impairing operator efficiency or creating significant discomfort • Physical distress to occupants of an aircraft (except operator), including injuries • Major occupational illness and/or major environmental damage, and/or major property damage
Minor	Does not significantly reduce system safety. Actions required by operators are well within their capabilities. Include • Slight reduction in safety margin or functional capabilities • Slight increase in workload, such as routine flight plan changes • Some physical discomfort to occupants or aircraft (except operators) • Minor occupational illness and/or minor environmental damage and/or minor property damage
No safety effect	Has no effect on safety

LIKELIHOOD OF OCCURRENCE

Likelihood	Definition
Probable	• Qualitative: Anticipated to occur one or more times during the entire system of the operational life of an item. • Quantitative: Probability of occurrence per operational hour is greater than 1×10^5 {\displaystyle 1\times 10^{5}}
Remote	• Qualitative: Unlikely to occur to each item during its total life. May occur several times in the life of an entire system or fleet. • Quantitative: Probability of occurrence per operational hour is less than the 1×10^5 {\displaystyle 1\times 10^{greater occurrence. 1×10^7 {\displaystyle 1\times 10^{7}}
Extremely remote	• Qualitative: not anticipated to occur to each item during its total life. May occur a few times in the life of an entire system or fleet. • Quantitative: probability of occurrence per operational hour is less than the 1×10^7 {\displaystyle 1\times 10^{greater than occurrence. 1×10^9 {\displaystyle 1\times 10^{9}}
Extremely improbable	• Qualitative: so unlikely that it is not anticipated to occur during the entire operational life of an entire system or fleet. • Quantitative: probability of occurrence per operational hour is less than occurrence. 1×10^9 {\displaystyle 1\times 10^{9}}

Again, hazard analysis is used as the first step in a process used to assess risk. The result of a hazard analysis is the identification of different type of hazards. A hazard is a potential condition and exists or not (probability is 1 or 0). It may be in single existence or in combination with other hazards (sometimes called events) and conditions become an actual functional failure or accident (mishap).

7 Composite Risk Management Process

The U.S. Army uses this process to lessen the impact of risk and threat on its operations, and I believe this process would be a valuable tool in the emergency management planning process.

The composite risk management (CRM) process is a process for decision making developed by the U.S. military to acknowledge, assess, and address hazards and control risks—during missions, operations, and even day-to-day activities (U.S. Army, 2010).

FIVE-STEP CRM PROCESS

- Step 1: Identify hazards to the force. Consider all aspects of current and future situations, environments, and known historical problem areas.
- Step 2: Assess hazards to determine risks. Assess the impact of each hazard in terms of potential loss and cost based on probability and severity.
- Step 3: Develop controls and make risk decisions. Develop control measures that eliminate the hazard or reduce its risk. As control measures are developed, risks are reevaluated until the residual risk is at a level where the benefits outweigh the cost. The appropriate decision authority then makes the decision.
- Step 4: Implement controls that eliminate the hazards or reduce their risks. Ensure the controls are communicated to all involved.
- Step 5: Supervise and evaluate. Enforce standards and controls. Evaluate the effectiveness of controls and adjust/update as necessary. Ensure lessons learned are fed back into the system for future planning.

LEVELS OF RISK MANAGEMENT

CRM is designed to help you in your decision-making process. It's not intended to be a time-consuming effort. Therefore, use only the amount of risk management necessary for the task. There are three levels of risk management:

1. *Time critical*: Used when there is little time, little complexity, or low risk. Often used during the execution phase of an operation where an unplanned change occurs and must be managed. It's easily applied to any situations.
2. *Deliberate*: Used in the majority of workplace applications where experience in a group setting will produce the best results.
3. *Strategic*: Used in high-priority or high-visibility situations, strategic CRM generally requires use of more thorough hazard identification and risk assessment tools. It is generally reserved for the more complex and riskier efforts, as it may be time-consuming.

CRM PRINCIPLES

CRM has four key principles that govern its application. These underlying principles should be considered before, during, and after every application of the five-step process.

- *Accept no unnecessary risk.* Of course, all task and daily routines involve risk. You must accept necessary risk required to successfully complete the mission or task. Unnecessary risk comes without a corresponding return, in terms of real benefits or available opportunities. CRM is dedicated to exposing and avoiding unnecessary risks.
- *Make risk decisions at the appropriate level.* Those accountable for the success or failure of the mission must be included in the risk-decision process. This establishes clear accountability. Commanders must ensure that subordinates know how much risk they may accept and when they must elevate the decision to a higher level.
- *Accept the risk when the benefits outweigh the costs.* Even high-risk endeavors may be undertaken when there is a clear knowledge that the sum of the benefits exceeds the sum of the costs. CRM is about controlling risk, not avoiding all risk. Integrate CRM into Doctrine and Planning at all Levels.
- *Integrating risk management into planning as early as possible provides the decision-maker the greatest opportunity to apply CRM principles.* Usually, it reduces costs and enhances CRM's overall effectiveness too.

RISK ASSESSMENT MATRIX

Once exposure, severity, and probability have been determined, you can now complete the last action of risk assessment. Combine the severity and probability estimates to form a Risk Assessment Matrix as shown in Figure 7.1. For example, if you have a hazard whose severity is judged to be critical but is unlikely to happen, the risk level is low.

Risk assessment matrix						
		Probability				
Severity		Frequent A	Likely B	Occasional C	Seldom D	Unlikely E
Catastrophic	I	E	E	H	H	M
Critical	II	E	H	H	M	L
Marginal	III	H	M	M	L	L
Negligible	IV	M	L	L	L	L
E – extremely high		H – high		M – moderate		L – low

FIGURE 7.1 Risk assessment matrix.

Remember that the risk assessment matrix doesn't directly consider exposure levels. You must factor any exposure considerations into the judgment of severity and probability. For example, a normally seldom occurrence of equipment failure may be increased to occasional if the exposure rate is high, thus making the overall risk level medium.

RISK PRIORITY LIST

The output of risk assessment is a prioritization of risk. You need to list the risks in order from the most serious risk to the least serious risk of any consequence. This ensures that controls are first identified for the most serious threat to mission accomplishment. All steps in the risk assessment process should be fully documented for future reference.

COMPONENTS OF RISK

EXPOSURE

Exposure is the first action in assessing risk. Remember that exposure is the number of resources (personnel or equipment) affected by a given event or by repeated events over time. Exposure can be expressed in the following terms:

- *Time*—how long resources are exposed to the hazard.
- *Proximity*—how close to resources would the hazard occur.
- *Volume*—the number of resources affected by the hazard.
- *Repetition*—the frequency with which the hazard would recur.

Although exposure is a component of risk, it is not used directly in assigning a level of risk. However, you should consider the exposure potential when assigning levels of probability and severity.

SEVERITY

The severity assessment should be based on the worst possible outcome that can be expected. It's expressed in terms of the impact on mission, people, material, facilities, and environment. Rank hazard severity using the following standard categories:

PROBABILITY

Next, estimate the probability of the hazard. Probability tells you how often the hazard will affect some operation within the mission. If you have reliable data available, you can also express probability as a number, in terms of a ratio or as a percentage. Express the level of probability by using the following ranking levels.

Keep in mind the cumulative probability of the causes listed for the hazard. For example, if there are four causes for a single hazard, the probability of its occurrence will be greater than if there were only one cause. For future reference, document your supporting rationale for assigning a probability to each hazard.

- *Frequent*—Occurs often, continuously experienced
- *Likely*—Occurs several times
- *Occasional*—Occurs sporadically
- *Seldom*—Unlikely, but could occur at some time
- *Unlikely*—Can assume it will not

Figure 7.1 presents a simple chart to assist one in the CRM process. This chart can be made up of whatever means one needs that may apply to one risk and threat, kind of fill in the blanks to be adaptable to one's needs.

8 Hazards (Risk)

A hazard is any biological, chemical, mechanical, environmental, or physical agent that is reasonably likely to cause harm or damage to humans, other organisms, or the environment in the absence of its control. This can include, but is not limited to, asbestos, electricity, microbial pathogens, motor vehicles, nuclear power plants, pesticides, vaccines, and X-rays. Identification of hazards is the first step in performing a risk assessment, although in some cases, risk assessment may not be necessary.

HAZARD TYPES

Biological hazards are associated with food, including certain viruses, parasites, fungi, bacteria, and plant and seafood toxins. Pathogenic Campylobacter and Salmonella are common food-borne biological hazards. The hazards from these bacteria can be avoided through risk mitigation steps such as proper handling, storing, and cooking of food. Disease in humans can come from biological hazards in the form of infection by bacteria, antigens, viruses, or parasite. There is some concern that new technologies such as genetic engineering pose biological hazards. Genetically modified (GM) organisms are relatively new manmade biological hazards and many have yet to be fully characterized.

Chemicals can be considered a hazard if, by virtue of its intrinsic properties, it can cause harm or danger to humans, property, or the environment.

Some harmful chemicals occur naturally in certain geological formations, such as radon gas or arsenic. Other chemicals include products with commercial uses, such as agricultural and industrial chemicals, as well as products developed for home use. Pesticides, which are normally used to control unwanted insects and plants, may cause a variety of negative effects on nontarget organisms.

Dichloro-diphenyl-trichloroethane (DDT) can build up, or bioaccumulate, in birds, resulting in thinner-than-normal egg shells, which can break in the nest. The organo chlorine pesticide dieldrin has been linked to Parkinson's disease. Corrosive chemicals like sulfuric acid, which is found in car batteries and research laboratories, can cause severe skin burns. Many other chemicals used in industrial and laboratory settings can cause respiratory, digestive, or nervous system problems if they are inhaled, ingested, or absorbed through the skin. The negative effects of other chemicals, such as alcohol and nicotine, have been well documented. Hazards associated with chemicals are dependent on the dose or amount of the chemical.

MECHANICAL

A mechanical hazard is any hazard involving a machine or process. Motor vehicles, aircraft, and air bags pose mechanical hazards. Compressed gases or liquids can also be considered a mechanical hazard.

Physical

A physical hazard is a naturally occurring process that has the potential to create loss or damage. Physical hazards include earthquakes, floods, and tornadoes. Physical hazards often have both human and natural elements. Flood problems can be affected by the natural elements of climate fluctuations and storm frequency and by land drainage and building in a flood plain, human elements.

Another physical hazard, X-rays, naturally occurs from solar radiation but has also been utilized by humans for medical purposes; however, overexposure can lead to cancer, skin burns, and tissue damage.

HAZARD VS RISK

The terms "hazard" and "risk" are often used interchangeably. However, in terms of risk assessment, these are two very distinct terms. As defined above, a hazard is any biological, chemical, mechanical, or physical agent that is reasonably likely to cause harm or damage to humans or the environment with sufficient exposure or dose. Risk is defined as the probability that exposure to a hazard will lead to a negative consequence or, more simply, Risk = Hazard × Dose (Exposure). Thus, a hazard poses no risk if there is not exposure to that hazard.

HAZARD IDENTIFICATION

Mechanical and Physical Hazards

Many mechanical hazards (aircraft, motor vehicles) and physical hazards (earthquakes, floods) have already been identified and well described. Hazard identification of new machines and/or industrial processes occurs at various stages in the design of the new machine or process. These hazard identification studies focus mainly on deviations from the intended use or design and the harm that may occur as a result of these deviations. These studies are regulated by various agencies such as the Occupational Safety and Health Administration and the National Highway Traffic Safety Administration.

Biological Hazards

Many biological hazards have also been identified. For example, the hazards of naturally occurring bacteria such as *Escherichia coli* and Salmonella are well known as disease-causing pathogens, and a variety of measures have been taken to limit human exposure to these microorganisms through food safety, good personal hygiene, and education. However, the potential for new biological hazards exists through the discovery of new microorganisms and through the development of new GM organisms. Use of new GM organisms is regulated by various governmental agencies. The US Environmental Protection Agency (EPA) controls GM plants that produce or resist pesticides (i.e., BT corn and Roundup ready crops). The US Food and Drug Administration (FDA) regulates GM plants that will be used as food or for medicinal purposes.

Chemical Hazards

A variety of chemical hazards (DDT, atrazine) have been identified as well. However, every year, companies produce more new chemicals to fill new needs or to take the place of older, less effective chemicals. Laws, such as the Federal Food, Drug, and Cosmetic Act and the Toxic Substances Control Act in the United States, require protection of human health and the environment for any new chemical introduced. In the United States, the EPA regulates new chemicals that may have environmental impacts (i.e., pesticides or chemicals released during a manufacturing process), while the FDA regulates new chemicals used in foods or as drugs. The potential hazards of these chemicals can be identified by performing a variety of tests prior to the authorization of usage. The amount of tests required and the extent to which the chemicals are tested vary, depending on the desired usage of the chemical. Chemicals designed as new drugs must undergo more rigorous tests than those used as pesticides.

Environmental Hazards

Natural Hazards

Natural hazard risks have threatened people, society, the natural environment, and the built environment, particularly more vulnerable people, throughout history and, in some cases, on a day-to-day basis. The social, natural, and built environments are at risk not only from geophysical hazards such as earthquakes, floods, volcanoes, and tsunami but also from manmade technological hazards, including industrial explosions, release of chemical hazards, and major accident hazards. Manmade hazards include the emergence of risks to people and the built environment in the modern world where hazards are presented by terrorist threats and technological hazards. According to the Red Cross (Alexander, 2014), each year, 130,000 people are killed, 90,000 are injured, and 140 million are affected by unique events known as disaster, and the effects of the loss of life, property damage, and social and economic disruption are caused by natural disasters, such as earthquakes, wind storms, tsunamis, floods, landslides, volcanic eruptions, wildfire, grasshopper and locust infestations, drought, and desertification and other calamities of natural origin.

Mitigating Natural Hazards

Methods to reduce risk from natural hazards include construction of prone facilities away from areas with high risk, redundancy, emergency reserve funds, purchasing relevant insurance, and the development of operational recovery plans.

Natural Hazard and Disaster Definitions

Disaster can be defined as the degree of risk of hazard that has a defined potential to cause significant personal, societal, property, or environmental damage or destruction. Disaster can manifest in various forms, threatening those people or environments specifically vulnerable to disaster. For example, a hurricane making landfall in the southeast United States presents a risk to those people, buildings, and environments in the path of the hurricane and within proximity to the hazard where there is a risk and potentially those who will be involved in the emergency response. Such impacts include death, injury, trauma, or posttraumatic stress disorder.

The definition of an environmental hazard is "the threat potential posed to man or nature by events originating in, or transmitted by, the natural or built environment." Keith Smith (2001) says that this definition includes a broader range of hazards ranging from long-term environmental deterioration, such as acidification of soils and build-up of atmospheric carbon dioxide, to communal and involuntary social hazards such as crime and terrorism to voluntary and personal hazards such as drug abuse and mountain climbing.

Environmental hazards usually have defined or common characteristics, including their tendency to be rapid-onset events, meaning they occur with a short warning time, they have a clear source of origin which is easily identified, impact will be swift and losses are suffered quickly during or shortly after the onset of the event, risk of exposure is usually involuntary due to location or proximity of people to set hazard, and the disaster occurs with an intensity and scale that justify an emergency response.

Hazards were grouped by Christopher and Burton (2001) according to their characteristics. These were factors related to geophysical events which were not process specific (Smith, 2001). They were the following:

1. Areal extent of damage zone
2. Intensity of impact at a point
3. Duration of impact at a point
4. Rate of onset of the event
5. Predictability of the event

Disaster can take various forms, including hurricane, volcano, tsunami, earthquake, drought, famine, plague, disease, rail crash, car crash, tornado, deforestation, flooding, toxic release, and spills.

These can affect people and the environment on the local regional level, national level, or international level, where the international community becomes involved with aid donation and governments give money to support affected countries' economies with disaster response and reconstruction postdisaster.

In defining hazard, it is important to distinguish between natural hazards, which may be defined as extreme events that originate in the biosphere, hydrosphere, lithosphere or atmosphere and a potential threat to humans and their welfare, which include earthquakes, landslides, hurricanes, and tsunamis, and technological hazards, including explosions, release of toxic materials, episodes of severe contamination, structural collapses, and transportation, construction and manufacturing accidents, etc. There is also a distinction to be made between rapid-onset natural hazards, technological hazards, and social hazards and the consequences of environmental degradation such as desertification and drought, which are described as being of sudden occurrence and relatively short duration. The distinction between hazard and risk are defined as the hazard will cause harm or damage and the chance of the hazard happening is not mitigated. Risk has the additional implication of the chance of a particular hazard actually occurring and thus define risk as the probability of hazard occurrence. Major disaster, as it is usually assessed on quantitative criteria of death and damage, was defined by Hewitt and Sheehan (1969) as having to conform to the following criteria:

- At least 100 people dead
- At least 100 people injured
- At least $1 million damage

RISK

Risk can be defined as the likelihood or probability of a given hazard of a given level causing a particular level of loss of damage. The elements of risk to the populations and communities come from the built-up environment, the natural environment, and economic activities and services which are under threat of disaster in any given area. Risk can be equated with a simple equation, although it is not mathematical. The total risk according to UNDRO (1982) is the "sum of predictable deaths, injuries, destruction, damage, disruption, and costs of repair and mitigation caused by a disaster of a particular level in a given area or areas." Mathematically, it can be written as

$$\text{Total risk} = (\text{Sum of the elements at risk}) \times (\text{Hazard} \times \text{Vulnerability})$$

Distinguishing between risk and vulnerability, we can say that vulnerability refers to the potential for casualty, destruction, damage, disruption or other form of loss in a particular element: risk combines this with the probable level of loss to be expected from a predictable magnitude of hazard. As hazard has varying degrees of severity, the more intense or severe the hazard, the greater vulnerability there will be as potential for damage and destruction is increased with respect to severity of hazard. Ben Wisner (2003) argues that risk or disaster is "a compound function of the natural hazard and the number of people, characterized by their varying degrees of vulnerability to that specific hazard, who occupy the space and time of exposure to the hazard event." This is simplified into a nonmathematical equation: $R = H \times V$.

Risk, vulnerability, and hazard are the three factors or elements that we are considering here in this pseudo-equation. Another definition of risk given by factor analysis of information risk that may be related to disaster is "the probable frequency and probable magnitude of future losses." Again, this definition focuses on the probability of future loss whereby the degree of vulnerability to hazard represents the level of risk on a particular population, built environment, or environment. The relationship between severity of environmental hazard, the probability of it happening, and risk will obviously vary, and it is necessary to outline the threats posed by a hazard. These are the following:

1. Hazards to people—death, injury, disease, and stress
2. Hazards to goods—property damage and economic loss
3. Hazards to environment—loss of flora and fauna, pollution, and loss of amenity

SMAUG MODEL—A BASIS FOR PRIORITIZING HAZARD RISKS

In emergency or disaster management, the SMAUG model of identifying and prioritizing risk of hazards associated with natural and technological threats is an effective tool. SMAUG stands for seriousness, manageability, acceptability, urgency, and growth, which are the criteria used for prioritization of hazard risks. The SMAUG model provides an effective means of prioritizing hazard risks based upon the aforementioned criteria in order to address the risks posed by the hazards to the avail of effecting effective mitigation, reduction, and response and recovery methods.

- *Seriousness* can be defined as "the relative impact in terms of people and dollars." This includes the potential for lives to be lost and potential for injury, as well as the physical, social, and as mentioned, economic losses that may be incurred
- *Manageability* can be defined as "the relative ability to mitigate or reduce the hazard (through managing the hazard, or the community or both)." Hazards presenting a high risk and as such requiring significant amounts of risk reduction initiatives will be rated high.
- *Acceptability* is the degree to which the risk of hazard is acceptable in terms of political, environmental, social, and economic impact
- *Urgency* is related to the probability of risk of hazard and is defined in terms of how imperative it is to address the hazard
- *Growth* is the potential for the hazard or event to expand or increase in either probability or risk to community or both. Should vulnerability increase, potential for growth may also increase.

An example of the numerical ratings for each of the four criteria is shown in the following:

- Manageability: high = 7+, medium = 5–7, low = 0–4
- Urgency: high = >20 years, medium = <20 years, low = 100 years
- Acceptability: high priority—poses more significant risk; low priority—lower risk of hazard impact
- Growth: high = 3, medium = 2, low = 1
- Seriousness: high = 4–5, medium = 2–3, low = 0–1

HIERARCHY OF HAZARD CONTROL

Hierarchy of hazard control is a system used in industry to minimize or eliminate exposure to hazards (Figure 8.1). It is a widely accepted system promoted by numerous safety organizations. This concept is taught to managers in industry, to be promoted as standard practice in the workplace. Various illustrations are used to depict this system, most commonly a triangle.

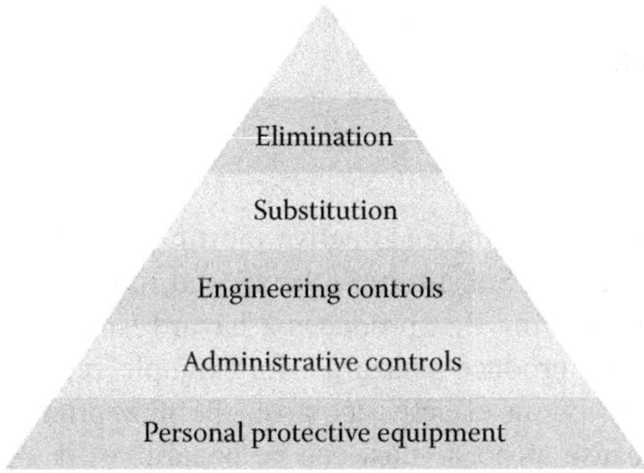

FIGURE 8.1 Hierarchy of hazard control.

The hazard controls in the hierarchy are, in order of decreasing effectiveness:

- Elimination
- Substitution
- Engineering
- Administration
- Personal protective equipment (PPE)

Components of the Hierarchy

The components of the hierarchy are shown in Figure 8.2.

Elimination

Eliminating the hazard—physically removing it is the most effective hazard control. For example, if employees must work high above the ground, the hazard can be

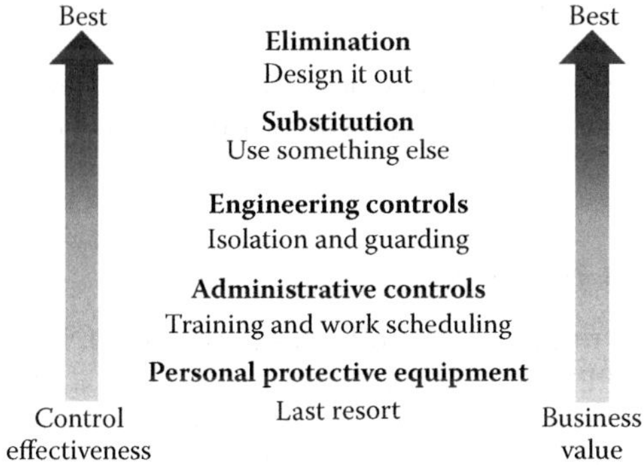

FIGURE 8.2 Components of the hierarchy.

eliminated by moving the piece they are working on to ground level to eliminate the need to work at heights.

Substitution

This pesticide contains DDT, and an effective substitution would be to replace it with a environmentally friendly. Substitution, the second most effective hazard control, involves replacing something that produces a hazard (similar to elimination) with something that does not produce a hazard—for example, replacing lead-based paint with acrylic paint. To be an effective control, the new product must not produce another hazard. Because airborne dust can be hazardous, if a product can be purchased with a larger particle size, the smaller product may effectively be substituted with the larger product.

Engineering Controls

The third most effective means of controlling hazards is engineered controls. These do not eliminate hazards, but rather isolate people from hazards. Capital costs of engineered controls tend to be higher than less effective controls in the hierarchy; however, they may reduce future costs. For example, a crew might build a work platform rather than purchase, replace, and maintain fall arrest equipment. "Enclosure and isolation" create a physical barrier between personnel and hazards, such as using remotely controlled equipment. Fume hoods can remove airborne contaminants as a means of engineered control.

Administrative Controls

Administrative controls are changes to the way people work. Examples of administrative controls include procedure changes, employee training, and installation of signs and warning labels (such as those in the Workplace Hazardous Materials Information System [WHMIS], https://www.canada.ca/en/health-canada/services/environmental).

Administrative controls do not remove hazards, but limit or prevent people's exposure to the hazards, such as completing road construction at night when fewer people are driving.

Personal Protective Equipment

PPE includes gloves, respirators, hard hats, safety glasses, high-visibility clothing, and safety footwear. PPE is the least effective means of controlling hazards because of the high potential for damage to render PPE ineffective. Additionally, some PPE, such as respirators, increase physiological effort to complete a task and therefore may require medical examinations to ensure that workers can use the PPE without risking their health.

9 Vulnerability

Vulnerability refers to the inability (of a system or a unit) to withstand the effects of a hostile environment. A *window of vulnerability* is a time frame within which defensive measures are diminished, compromised, or lacking.

COMMON APPLICATIONS

In relation to hazards and disasters, vulnerability is a concept that links the relationship that people have with their environment to social forces and institutions and the cultural values that sustain and contest them. "The concept of vulnerability expresses the multi dimensionality of disasters by focusing attention on the totality of relationships in a given social situation which constitute a condition that, in combination with environmental forces, produces a disaster" (Bankoff, 2004).

It's also the extent to which changes could harm a system, or to which the community can be affected by the impact of a hazard or exposed to the possibility of being attacked or harmed, either physically or emotionally: "we were in a vulnerable position."

RESEARCH

Within the body of literature related to vulnerability, major research streams include questions of methodology, such as measuring and assessing vulnerability, including finding appropriate indicators for various aspects of vulnerability, up and down scaling methods, and participatory methods.

Vulnerability research covers a complex, multidisciplinary field including development and poverty studies, public health, climate studies, security studies, engineering, geography, political ecology, and disaster risk management. This research is of importance and interest for organizations trying to reduce vulnerability—especially as related to poverty and other Millennium Development Goals. Many institutions are conducting interdisciplinary research on vulnerability. A forum that brings many of the current researchers on vulnerability together is the Expert Working Group (Biometric Vulnerability Assessment Expert Group, https://www.biometricsinstitute.org/biometric-vulnerability). Researchers are currently working to refine definitions of "vulnerability," measurement and assessment methods, and effective communication of research to decision makers.

TYPES OF VULNERABILITIES

SOCIAL VULNERABILITY

In its sense, social vulnerability is one dimension of vulnerability to multiple stressors (agent responsible for stress) and shocks, including abuse, social exclusion, and natural hazards. Social vulnerability refers to the inability of people, organizations, and societies to withstand adverse impacts from multiple stressors to which they are exposed. These impacts are due in part to characteristics inherent in social interactions, institutions, and systems of cultural values.

Cognitive Vulnerability

A cognitive vulnerability, in cognitive psychology, is an erroneous belief, cognitive bias, or pattern of thought that is believed to predispose the individual to psychological problems.

It is in place before the symptoms of psychological disorders start to appear, such as high neuroticism, and after the individual encounters a stressful experience, the cognitive vulnerability shapes a maladaptive response that may lead to a psychological disorder. In psychopathology, cognitive vulnerability is constructed from schema models, hopelessness models, and attachment theory. Attentional bias is one mechanism leading to faulty cognitive bias that leads to cognitive vulnerability. Allocating a danger level to a threat depends on the urgency or intensity of the threshold. Anxiety is not associated with selective orientation.

Military

In military terminology, vulnerability is a subset of survivability, the others being susceptibility and recoverability. Vulnerability is defined in various ways depending on the nation and service arm concerned, but in general, it refers to the near-instantaneous effects of a weapon attack. In aviation, it is defined as the inability of an aircraft to withstand the damage caused by the manmade hostile environment. In some definitions, recoverability (damage control, firefighting, restoration of capability) is included in vulnerability. Some military services develop their own concept of vulnerability.

Invulnerability

Invulnerability is a common feature found in video games. It makes the player impervious to pain, damage, or loss of health. It can be found in the form of "power-ups" or cheats; when activated via cheats, it is often referred to as "god mode." Generally, it does not protect the player from certain instant-death hazards, most notably "bottomless" pits from which, even if the player were to survive the fall, they would be unable to escape.

As a rule, invulnerability granted by power-ups is temporary and wears off after a set amount of time, while invulnerability cheats, once activated, remain in effect until deactivated, or the end of the level is reached. Depending on the game in question, invulnerability to damage may or may not protect the player from nondamage effects, such as being immobilized or sent flying.

VULNERABILITY ASSESSMENT

A vulnerability assessment is the process of identifying, quantifying, and prioritizing (or ranking) the vulnerabilities in a system. Examples of systems for which vulnerability assessments are performed include, but are not limited to, information technology (IT) systems, energy supply systems, water supply systems, transportation systems, and communication systems.

Such assessments may be conducted on behalf of a range of different organizations, from small businesses up to large regional infrastructures. Vulnerability from the perspective of disaster management means assessing the threats from potential hazards to the population and to infrastructure. It may be conducted in the political, social, economic, or environmental fields.

Vulnerability assessment has many things in common with risk assessment. Assessments are typically performed according to the following steps:

- Cataloging assets and capabilities (resources) in a system
- Assigning quantifiable value (or at least rank order) and importance to those resources
- Identifying the vulnerabilities or potential threats to each resource
- Mitigating or eliminating the most serious vulnerabilities for the most valuable resources

Classical risk analysis is described as follows: "Classical risk analysis is principally concerned with investigating the risks surrounding a plant (or some other object), its design and operations. Such analysis tend to focus on causes and the direct consequences for the studied object. Vulnerability analysis, on the other hand, focus both on consequences for the object itself and on primary and secondary consequences for the surrounding environment. It also concerns itself with the possibilities of reducing such consequences and of improving the capacity to manage future incidents." In general, a vulnerability analysis serves to "categorize key assets and drive the risk management process" (Faulstich, 2015).

In the United States, guides providing valuable considerations and templates for completing a vulnerability assessment are available from numerous agencies, including the Department of Energy, the Environmental Protection Agency, and the U.S. Department of Transportation, just to name a few (Vulnerability Assessment Report report template, https://www.reporttemplates.net/.../Vulnerability_Assessment_Report).

The framework makes use of nested flowcharts to show how social and environmental forces interact to create situations vulnerable to sudden changes. Proposing an analytical framework, based on research, suggests that the first stage is to assess current vulnerability by documenting exposures and current adaptive strategies. This should be followed by a second stage that estimates directional changes in those current risk factors and characterizes the community's future adaptive capacity. The framework utilizes historic information including how communities have experienced and addressed climatic hazards, with information on what conditions are likely to change and what constraints and opportunities there are for future adaptation.

VULNERABILITY INDEX

A vulnerability index is a measure of the exposure of a population to some hazard. Typically, the index is a composite of multiple quantitative indicators that, via some formula, delivers a single numerical result. Through such an index, "diverse issues can be combined into a standardized framework…making comparisons possible" (SIDS of the Pacific, 2015). For instance, indicators from the physical sciences can be combined with social, medical, and even psychological variables to evaluate potential complications for disaster planning.

The origin of vulnerability indexes as a policy planning tool began with the United Nations Environmental Program. One of the participants in the early task forces has also conducted secondary research documenting the evolution of the analytic tool through various stages (United Nations Environment Programme, www.unep.org).

BASIC METHODOLOGY

The basic methodology of constructing a vulnerability index is that the individual measures are weighted according to their relative importance. A cumulative score is then generated, typically by adding the weighted values. Decision trees can evaluate alternative policy options.

IN HAZARD PLANNING

The concept has been extended and applied in dealing with risk from natural hazards and the part that population metrics play in making such a situation into a disaster. In the United States, this has been done at a county level and is run by the Hazards and Vulnerability Research Institute.

THREATS TO COMPUTERS AND IT SYSTEMS

In computer security, a threat is a possible danger that might exploit a vulnerability to breach security and therefore cause possible harm.

A threat can be either "intentional" (i.e., hacking: an individual cracker or a criminal organization) or "accidental" (e.g., the possibility of a computer malfunctioning or the possibility of a natural disaster such as an earthquake, a fire, or a tornado) or otherwise a circumstance, capability, action, or event.

Definitions

ISO 27005 defines threat as, "A potential cause of an incident that may result in harm of systems and organization" (ISO/IEC 27005:2011—Information technology—Security www.iso.org).

A more comprehensive definition, tied to an Information assurance point of view, can be found in "Federal Information Processing Standards (FIPS) 200 (Minimum Security Requirements for Federal, https://csrc.nist.gov/csrc/media/publications/fips/200/final/), Minimum Security Requirements for Federal Information and Information Systems" by the National Institute of Standards and Technology of the United States:

> Any circumstance or event with the potential to adversely impact organizational operations (including mission, functions, image, or reputation), organizational

assets, or individuals through an information system via unauthorized access, destruction, disclosure, modification of information, and/or denial of service. Also, the potential for a threat-source to successfully exploit a particular information system vulnerability.

The National Information Assurance Glossary defines threat as follows (https://www.ecs.csus.edu/csc/iac/cnssi_4009.pdf):

> Any circumstance or event with the potential to adversely impact an IS through unauthorized access, destruction, disclosure, modification of data, and/or denial of service.

European Union Agency for Network and Information Security gives a similar definition (https://www.enisa.europa.eu):

> Any circumstance or event with the potential to adversely impact an asset through unauthorized access, destruction, disclosure, modification of data, and/or denial of service.

The Open Group defines threat as follows:

> Anything that is capable of acting in a manner resulting in harm to an asset and/or organization; for example, acts of God (weather, geological events, etc.); malicious actors; errors; failures.

Factor analysis of information risk defines threat as follows:

> Threats are anything (e.g., object, substance, human, etc.) that are capable of acting against an asset in a manner that can result in harm. A tornado is a threat, as is a flood, as is a hacker. The key consideration is that threats apply the force (water, wind, exploit code, etc.) against an asset that can cause a loss event to occur.

The National Information Assurance Training and Education Center gives a more articulated definition of threat (niatec.info):

1. The means through which the ability or intent of a threat agent to adversely affect an automated system, facility, or operation can be manifest. Categorize and classify threats as follows: Categories Classes Human Intentional Unintentional Environmental Natural Fabricated.
2. Any circumstance or event with the potential to cause harm to a system in the form of destruction, disclosure, modification or data, and/or denial of service.
3. Any circumstance or event with the potential to cause harm to the ADP system or activity in the form of destruction, disclosure, and modification of data, or denial of service. A threat is a potential for harm. The presence of a threat does not mean that it will necessarily cause actual harm. Threats exist because of the very existence of the system or activity and not because of any specific weakness. For example, the threat of fire exists at all facilities regardless of the amount of fire protection available.

4. Types of computer systems related adverse events that may result in losses. Examples are flooding, sabotage, and fraud.
5. An assertion primarily concerning entities of the external environment (agents); we say that an agent (or class of agents) poses a threat to one or more assets; we write: T (e; i) where: e is an external entity; i is an internal entity or an empty set.
6. An undesirable occurrence that might be anticipated but is not the result of a conscious act or decision. In threat analysis, a threat is defined as an ordered pair, <peril; asset category>, suggesting the nature of these occurrences but not the details (details are specific to events).
7. A potential violation of security.
8. A set of properties of a specific external entity (which may be either an individual or class of entities) that, in union with a set of properties of a specific internal entity, implies a risk (according to some body of knowledge).

PHENOMENOLOGY

The term "threat" relates to some other basic security terms, as shown in Figure 9.1.

A resource (both physical and logical) can have one or more vulnerabilities that can be exploited by a threat agent in a threat action. The result can potentially compromise the confidentiality, integrity, or availability properties of resources (potentially different from the vulnerable one) of the organization and others involved parties (customers and suppliers).

The so-called Central Intelligence Agency (CIA) triad is the basis of information security.

The attack can be active when it attempts to alter system resources or affect their operation, so it compromises integrity or availability. A "passive attack" attempts to learn or make use of information from the system but does not affect system resources: so it compromises confidentiality.

OWASP: RELATIONSHIP BETWEEN THREAT AGENT AND BUSINESS IMPACT

Open Web Application Security Project (OWASP) depicts the same phenomenon in slightly different terms: a threat agent through an attack vector exploits a weakness

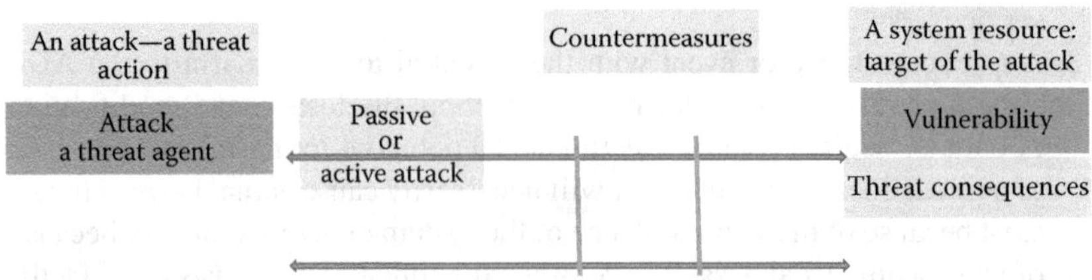

FIGURE 9.1 Threat flow chart.

(vulnerability) of the system and the related security controls, causing a technical impact on an IT resource (asset) connected to a business impact.

A set of policies concerned with information security management, the information security management systems (ISMS) (www.iso.org), has been developed to manage, according to risk management principles, the countermeasures in order to accomplish a security strategy set-up following rules and regulations applicable in a country. Countermeasures are also called security controls; when applied to the transmission of information, they are named security services.

The overall picture represents the risk factors of the risk scenario (Figure 9.2).

The widespread of computer dependencies and the consequent raising of the consequence of a successful attack, led to a new term: cyber warfare.

It should be noted that, nowadays, the many real attacks exploit psychology at least as much as technology. Phishing and pretexting and other methods are called social engineering techniques. The Web 2.0 applications, specifically social network services, can be a means to get in touch with people in charge of system administration or even system security, inducing them to reveal sensitive information.

The most widespread documentation on computer insecurity is about technical threats such as a computer virus, Trojan, and other malware, but a serious study to apply cost-effective countermeasures can only be conducted following a rigorous IT risk analysis in the framework of an ISMS: a pure technical approach will let out the psychological attacks that are increasing threats.

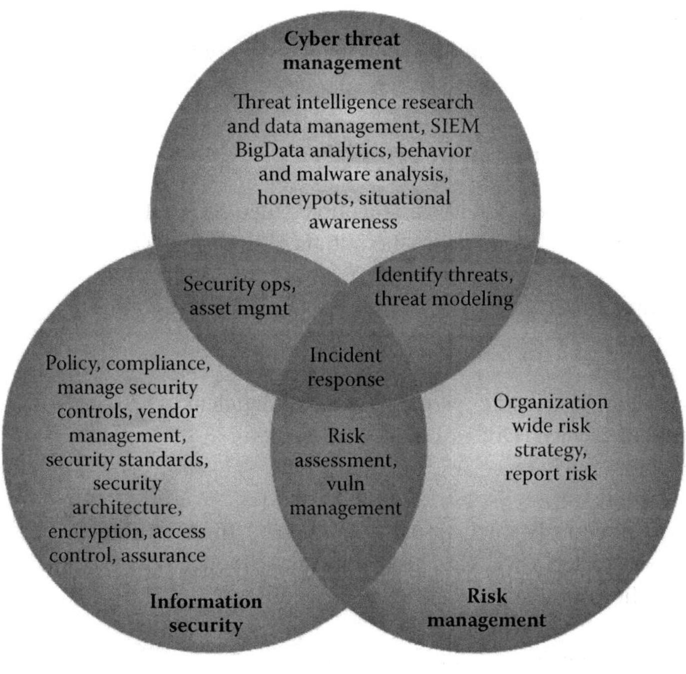

FIGURE 9.2 Risk factors of the risk scenario.

THREATS CLASSIFICATION

Threats can be classified according to their type and origin (Figure 9.3).

Types of threats:

- Physical damage: fire, water, and pollution
- Natural events: climatic, seismic, and volcanic
- Loss of essential services: electrical power, air conditioning, and telecommunication
- Compromise of information: eavesdropping, theft of media, and retrieval of discarded materials
- Technical failures: equipment, software, and capacity saturation
- Compromise of functions: error in use, abuse of rights, and denial of actions

Note that a threat type can have multiple origins.

- Deliberate: aiming at information asset
- Spying
- Illegal processing of data
- Accidental
- Equipment failure
- Software failure
- Environmental
- Natural event
- Loss of power supply
- Negligence: known but neglected factors, compromising the network safety and sustainability

SYSTEM THREAT CLASSIFICATION

Microsoft has proposed a threat classification called STRIDE (The STRIDE Threat Model, msdn.microsoft.com), from the initials of threat categories:

- Spoofing of user identity
- Tampering
- Repudiation
- Information disclosure (privacy breach or data leak)
- Denial of service (DoS)
- Elevation of privilege

Microsoft previously rated the risk of security threats using five categories in a classification called DREAD: Risk assessment model. The model is considered obsolete by Microsoft. The categories were as follows:

- *Damage*—how bad would an attack be?
- *Reproducibility*—how easy it is to reproduce the attack?

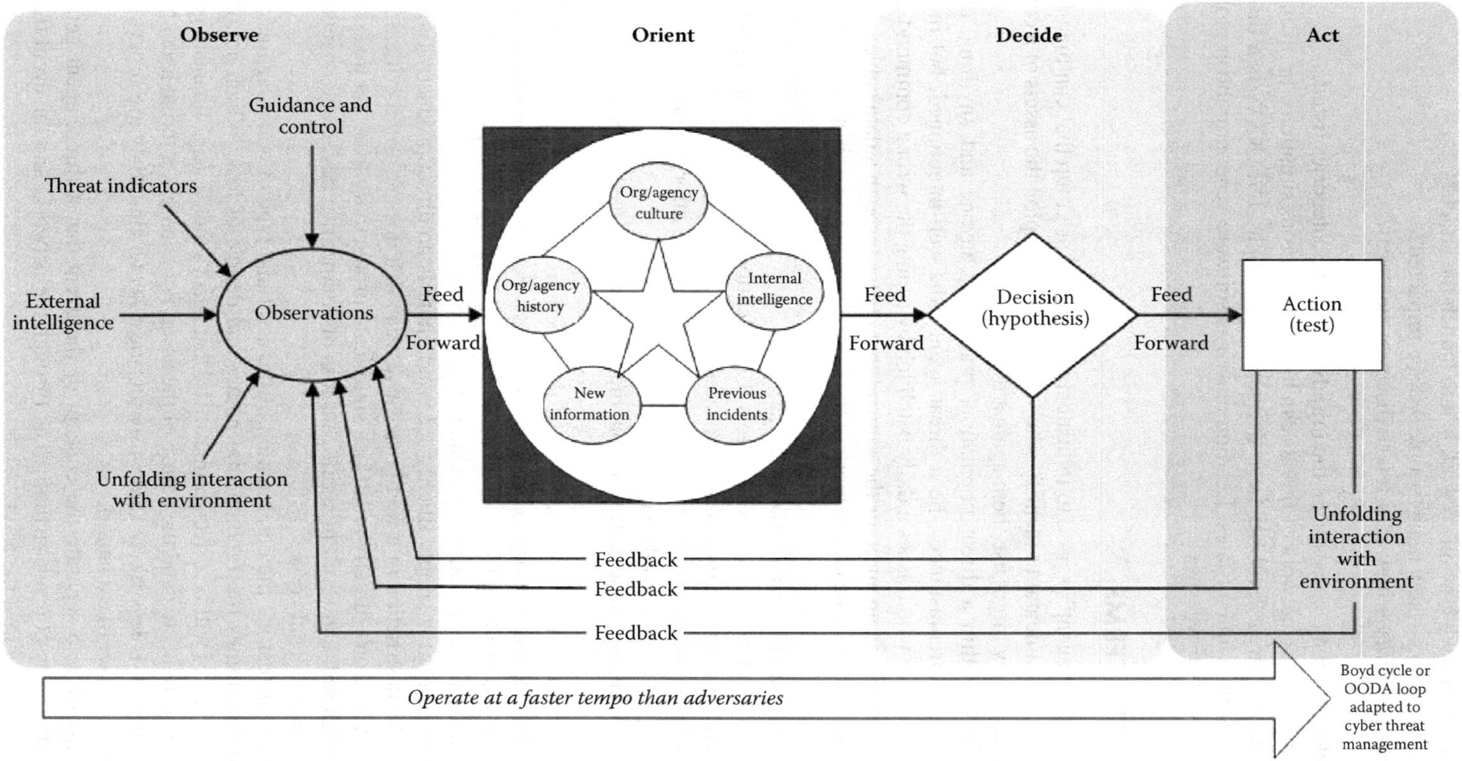

FIGURE 9.3 Threat classification decision chart.

- *Exploitability*—how much work is it to launch the attack?
- *Affected users*—how many people will be impacted?
- *Discoverability*—how easy it is to discover the threat?

The DREAD name comes from the initials of the five categories listed.

The spread over a network of threats can lead to dangerous situations. In military and civil fields, threat level has been defined: for example, INFOCON is a threat level used by the United States. Leading antivirus software vendors publish global threat level on their websites.

ASSOCIATED TERMS

The term "threat agent" is used to indicate an individual or group that can manifest a threat. It is fundamental to identify who would want to exploit the assets of a company, and how they might use them against the company.

Individuals within a threat population, practically anyone and anything, can, under the right circumstances, be a threat agent—the well-intentioned, but inept, computer operator who trashes a daily batch job by typing the wrong command, the regulator performing an audit, or the squirrel that chews through a data cable.

Threat agents can take one or more of the following actions against an asset:

- *Access*—simple unauthorized access
- *Misuse*—unauthorized use of assets (e.g., identity theft, setting up a porn distribution service on a compromised server, etc.)
- *Disclose*—the threat agent illicitly discloses sensitive information
- *Modify*—unauthorized changes to an asset
- *Deny access*—includes destruction, theft of a nondata asset, etc.

It's important to recognize that each of these actions affects different assets differently, which drives the degree and nature of loss. For example, the potential for productivity loss resulting from a destroyed or stolen asset depends upon how critical that asset is to the organization's productivity.

If a critical asset is simply illicitly accessed, there is no direct productivity loss. Similarly, the destruction of a highly sensitive asset that doesn't play a critical role in productivity won't directly result in a significant productivity loss. Yet that same asset, if disclosed, can result in significant loss of competitive advantage or reputation and generate legal costs.

The point is that it's the combination of the asset and type of action against the asset that determines the fundamental nature and degree of loss. Which action(s) a threat agent takes will be driven primarily by that agent's motive (e.g., financial gain, revenge, recreation, etc.) and the nature of the asset. For example, a threat agent bent on financial gain is less likely to destroy a critical server than they are to steal an easily pawned asset like a laptop.

It is important to separate the concept of the event that a threat agent gets in contact with the asset (even virtually, i.e., through the network) and the event that a threat agent acts against the asset.

OWASP collects a list of potential threat agents in order to prevent system designers and programmers insert vulnerabilities in the software.

- Threat Agent = Capabilities + Intentions + Past Activities

These individuals and groups can be classified as follows:

- Non-target specific: Non-target specific threat agents are computer viruses, worms, Trojans, and logic bombs.
- Employees: These include staff, contractors, operational/maintenance personnel, or security guards who are annoyed with the company.
- Organized crime and criminals: Criminals target information that is of value to them, such as bank accounts, credit cards, or intellectual property that can be converted into money. Criminals will often make use of insiders to help them.

Corporations

Corporations are engaged in offensive information warfare or competitive intelligence. Partners and competitors come under this category.

- Human, unintentional: accidents, carelessness
- Human, intentional: insider, outsider
- Natural: flood, fire, lightning, meteor, earthquakes

THREAT SOURCE

Threat sources are those who wish a compromise to occur. It is a term used to distinguish them from threat agents/actors who are those who actually carry out the attack and who may be commissioned or persuaded by the threat actor to knowingly or unknowingly carry out the attack.

Threat communities

Subsets of the overall threat agent population that share key characteristics, and the notion of threat communities is a powerful tool for understanding who and what we're up against as we try to manage risk.

For example, the probability that an organization would be subject to an attack from the terrorist threat community would depend, in large part, on the characteristics of your organization relative to the motives, intents, and capabilities of the terrorists.

1. Is the organization closely affiliated with ideology that conflicts with known, active terrorist groups?
2. Does the organization represent a high-profile, high-impact target?
3. Is the organization a soft target? How does the organization compare with other potential targets?

4. If the organization were to come under attack, what components of the organization would be likely targets?
5. For example, how likely is it that terrorists would target the company information or systems?

The following threat communities are examples of the human malicious threat landscape many organizations face:

Internal

- Employees
- Contractors (and vendors)
- Partners

External

- Cyber-criminals (professional hackers)
- Spies
- Nonprofessional hackers
- Activists
- Nation-state intelligence services (e.g., counterparts to the CIA, etc.)
- Malware (virus/worm/etc.) authors

THREAT ACTION

Threat action is an assault on system security. A complete security architecture deals with both intentional acts (i.e., attacks) and accidental events. Various kinds of threat actions are defined as subentries under "threat consequence."

THREAT ANALYSIS

Threat analysis is the analysis of the probability of occurrences and consequences of damaging actions to a system. It is the basis of risk analysis.

THREAT CONSEQUENCE

Threat consequence is a security violation that results from a threat action. It includes disclosure, deception, disruption, and usurpation.

The following subentries describe four kinds of threat consequences and also list and describe the kinds of threat actions that cause each consequence. Threat actions that are accidental events are marked by the following.

"Unauthorized disclosure (a threat consequence)": A circumstance or event whereby an entity gains access to data for which the entity is not authorized. (See: data confidentiality.) The following threat actions can cause unauthorized disclosure:

- "Exposure": A threat action whereby sensitive data is directly released to an unauthorized entity. This includes the following:

  - "Deliberate Exposure": Intentional release of sensitive data to an unauthorized entity.
  - "Scavenging": Searching through data residue in a system to gain unauthorized knowledge of sensitive data.
  - "Human error": Human action or inaction that unintentionally results in an entity gaining unauthorized knowledge of sensitive data.
  - "Hardware/software error": System failure that results in an entity gaining unauthorized knowledge of sensitive data.
- "Interception": A threat action whereby an unauthorized entity directly accesses sensitive data travelling between authorized sources and destinations. This includes:
  - "Theft": Gaining access to sensitive data by stealing a shipment of a physical medium, such as a magnetic tape or disk, that holds the data.
  - "Wiretapping (passive)": Monitoring and recording data that is flowing between two points in a communication system. (See: wiretapping.)
  - "Emanations analysis": Gaining direct knowledge of communicated data by monitoring and resolving a signal that is emitted by a system and that contains the data but is not intended to communicate the data.
- "Inference": A threat action whereby an unauthorized entity indirectly accesses sensitive data (but not necessarily the data contained in the communication) by reasoning from characteristics or byproducts of communications. This includes the following:
  - "Traffic analysis": Gaining knowledge of data by observing the characteristics of communications that carry the data.
  - "Signals analysis": Gaining indirect knowledge of communicated data by monitoring and analyzing a signal that is emitted by a system and that contains the data but is not intended to communicate the data. (See: Emanation.)
- "Intrusion": A threat action whereby an unauthorized entity gains access to sensitive data by circumventing a system's security protections. This includes the following:
  - "Trespass": Gaining unauthorized physical access to sensitive data by circumventing a system's protections.
  - "Penetration": Gaining unauthorized logical access to sensitive data by circumventing a system's protections.
  - "Reverse engineering": Acquiring sensitive data by disassembling and analyzing the design of a system component.
  - "Cryptanalysis": Transforming encrypted data into plain text without having prior knowledge of encryption parameters or processes.
- "Deception" (a threat consequence): A circumstance or event that may result in an authorized entity receiving false data and believing it to be true. The following threat actions can cause deception:
  - "Masquerade": A threat action whereby an unauthorized entity gains access to a system or performs a malicious act by posing as an authorized entity.
  - "Spoof": Attempt by an unauthorized entity to gain access to a system by posing as an authorized user.

    - "Malicious logic": In context of masquerade, any hardware, firmware, or software (e.g., Trojan horse) that appears to perform a useful or desirable function but actually gains unauthorized access to system resources or tricks a user into executing other malicious logic.
    - "Falsification": A threat action whereby false data deceives an authorized entity. (See: active wiretapping.)
    - "Substitution": Altering or replacing valid data with false data that serves to deceive an authorized entity.
    - "Insertion": Introducing false data that serve to deceive an authorized entity.
    - "Repudiation": A threat action whereby an entity deceives another by falsely denying responsibility for an act.
    - "False denial of origin": Action whereby the originator of data denies responsibility for its generation.
    - "False denial of receipt": Action whereby the recipient of data denies receiving and possessing the data.
- "Disruption" (a threat consequence): A circumstance or event that interrupts or prevents the correct operation of system services and functions. (See: denial of service.) The following threat actions can cause disruption:
    - "Incapacitation": A threat action that prevents or interrupts system operation by disabling a system component.
    - "Malicious logic": In the context of incapacitation, any hardware, firmware, or software (e.g., logic bomb) intentionally introduced into a system to destroy system functions or resources.
    - "Physical destruction": Deliberate destruction of a system component to interrupt or prevent system operation.
    - "Human error": Action or inaction that unintentionally disables a system component.
    - "Hardware or software error": Error that causes failure of a system component and leads to disruption of system operation.
    - "Natural disaster": Any natural disaster (e.g., fire, flood, earthquake, lightning, or wind) that disables a system component.
    - "Corruption": A threat action that undesirably alters system operation by adversely modifying system functions or data.
    - "Tamper": In the context of corruption, deliberate alteration of a system's logic, data, or control information to interrupt or prevent correct operation of system functions.
    - "Malicious logic": In the context of corruption, any hardware, firmware, or software (e.g., a computer virus) intentionally introduced into a system to modify system functions or data.
    - "Human error": Human action or inaction that unintentionally results in the alteration of system functions or data.
    - "Hardware or software error": Error that results in the alteration of system functions or data.
    - "Natural disaster": Any natural event (e.g., power surge caused by lightning) that alters system functions or data.

  - "Obstruction": A threat action that interrupts delivery of system services by hindering system operations.
  - "Interference": Disruption of system operations by blocking communications or user data or control information.
  - "Overload": Hindrance of system operation by placing excess burden on the performance capabilities of a system component. (See: flooding.)
- "Usurpation" (a threat consequence): A circumstance or event that results in control of system services or functions by an unauthorized entity. The following threat actions can cause usurpation:
  - "Misappropriation": A threat action whereby an entity assumes unauthorized logical or physical control of a system resource.
  - "Theft of service": Unauthorized use of service by an entity.
  - "Theft of functionality": Unauthorized acquisition of actual hardware, software, or firmware of a system component.
  - "Theft of data": Unauthorized acquisition and use of data.
  - "Misuse": A threat action that causes a system component to perform a function or service that is detrimental to system security.
  - "Tamper": In the context of misuse, deliberate alteration of a system's logic, data, or control information to cause the system to perform unauthorized functions or services.
  - "Malicious logic": In the context of misuse, any hardware, software, or firmware intentionally introduced into a system to perform or control execution of an unauthorized function or service.
  - "Violation of permissions": Action by an entity that exceeds the entity's system privileges by executing an unauthorized function.

THREAT LANDSCAPE OR ENVIRONMENT

A collection of threats in a particular domain or context along with the information can help identify vulnerable assets, threats, risks, threat actors, and observed trends.

THREAT MANAGEMENT

Threats should be managed by operating an information security management system (ISMS), performing all the IT risk management activities foreseen by laws, standards, and methodologies.

Very large organizations tend to adopt business continuity management plans in order to protect, maintain, and recover business-critical processes and systems. Some of these plans are foreseen to set up a computer security incident response team or a computer emergency response team.

There are two kinds of verification of the threat management process:

1. Information security audit
2. Penetration test

Most organizations perform a subset of these steps, adopting countermeasures based on a nonsystematic approach: computer insecurity studies the battlefield of computer security exploits and defenses that result.

Information security awareness is a significant market. There has been a lot of software developed to deal with IT threats, including both open-source software and proprietary software.

CYBER THREAT MANAGEMENT

Threat management involves a wide variety of threats, including physical threats like flood and fire. While the ISMS risk assessment process does incorporate threat management for cyber threats such as remote buffer overflows, the risk assessment process doesn't include processes such as threat intelligence management or response procedures.

Cyber threat management (CTM) is emerging as best practice for managing cyber threats beyond the basic risk assessment found in ISMS. It enables early identification of threats, data-driven situational awareness, accurate decision making, and timely threat mitigating actions.

CTM includes the following:

- Manual and automated intelligence gathering and threat analytics
- Comprehensive methodology for real-time monitoring including advanced techniques such as behavioral modeling
- Use of advanced analytics to optimize intelligence, generate security intelligence, and provide situational awareness
- Technology and skilled people leveraging situational awareness to enable rapid decisions and automated or manual actions

THREAT HUNTING

Cyber threat hunting is the process of proactively and iteratively searching through networks to detect and isolate advanced threats that evade existing security solutions. This is in contrast to traditional threat management measures, such as firewall intrusion detection systems, and security information and event management, which typically involve an investigation after there has been a warning of a potential threat or an incident has occurred.

Threat hunting can be a manual process, in which security analysts sift through various data information using their own knowledge and familiarity with the network to create hypotheses about potential threats. To be even more effective and efficient, however, threat hunting can be partially automated, or machine assisted, as well. In this case, the analyst utilizes a software that harnesses machine learning and user and entity behavior analytics (UEBA) to inform the analyst of potential risks.

The analyst then investigates these potential risks, tracking suspicious behavior in the network. Thus, hunting is an iterative process, meaning that it must be continuously carried out in a loop, beginning with a hypothesis. There are three types of hypotheses (Géron, 2017):

1. Analytics driven: "Machine-learning and UEBA, used to develop aggregated risk scores that can also serve as hunting hypotheses"
2. Situational-awareness driven: "Crown Jewel analysis, enterprise risk assessments, company- or employee-level trends"
3. Intelligence driven: "Threat intelligence reports, threat intelligence feeds, malware analysis, vulnerability scans"

Analysts research their hypothesis by going through vast amounts of data about the network. The results are then stored so that they can be used to improve the automated portion of the detection system and to serve as a foundation for future hypotheses.

Representative vendors of threat hunting software and services include the following:

- Carbon Black
- Domain tools
- Exabeam
- Raytheon Foreground Security
- Sqrrl
- Tenable Network Security

In closing, as mentioned in the first paragraph, computer security as a threat is a possible danger that might exploit a vulnerability to breach security and therefore cause possible harm.

A threat can be either "intentional" (i.e., hacking: an individual cracker or a criminal organization) or "accidental" (e.g., the possibility of a computer malfunctioning or the possibility of a natural disaster such as an earthquake, a fire, or a tornado) or otherwise a circumstance, capability, action, or event.

10 Emergency Management and Understanding the Impact of Terrorism

Terrorism is, in its broadest sense, the use of intentionally indiscriminate violence (terror) in order to achieve a political, religious, or ideological aim. It is classified as fourth-generation warfare and as a violent crime. In modern times, terrorism is considered a major threat to society and therefore illegal under antiterrorism laws in most jurisdictions. It is also considered a war crime under the laws of war when used to target noncombatants, such as civilians, neutral military personnel, or enemy prisoners of war.

A broad array of political organizations have practiced terrorism to further their objectives. It has been practiced by both right-wing and left-wing political organizations, nationalist groups, religious groups, revolutionaries, and ruling governments. The symbolism of terrorism can exploit human fear to help achieve these goals.

According to data from the Global Terrorism Database (http://www.start.umd.edu/gtd), more than 61,000 incidents of nonstate terrorism claiming over 140,000 lives have been recorded from 2000 to 2014.

ORIGIN OF THE TERM

"Terrorism" comes from the French word terrorism and originally referred specifically to state terrorism as practiced by the French government during the 1793–1794 Reign of Terror.

The French word terrorism in turn derives from the Latin verb terrere (e, terreo), meaning "to frighten." The terror cimbricus was a panic and state of emergency in Rome in response to the approach of warriors of the Cimbri tribe in 105 BCE that the Jacobins cited as a precedent when imposing the Reign of Terror during the French Revolution. After the Jacobins lost power, the word "terrorist" became a term of abuse.

Although terrorism originally referred to acts committed by a government, currently, it usually refers to the killing of innocent people for political purposes in such a way as to create a spectacle. This meaning can be traced back to Sergey Nechayev, who described himself as a terrorist. Nechayev founded the Russian terrorist group "People's Retribution" (Народная расправа) in 1869.

The lack of consensus as to what a terrorist is can affect policies designed to deal with terrorists. Some view them as soldiers who can be held at the end of a war and are entitled to various privileges spelled out in the Geneva Conventions. Others view them as criminals that should be tried in civil courts. Still, others will argue that terrorists are best treated as a category to themselves and need policies tailored to them.

In November 2004, a Secretary-General of the United Nations report described terrorism as any act "intended to cause death or serious bodily harm to civilians or non-combatants with the purpose of intimidating a population or compelling a government or an international organization to do or abstain from doing any act."

DEFINITION OF TERRORISM

The definition of terrorism has proven controversial. Various legal systems and government agencies use different definitions of terrorism in their national legislation. Moreover, the international community has been slow to formulate a universally agreed, legally binding definition of this crime. These difficulties arise from the fact that the term "terrorism" is politically and emotionally charged.

The international community has never succeeded in developing an accepted comprehensive definition of terrorism. During the 1970s and 1980s, the United Nations' attempts to define the term floundered mainly due to differences of opinion between various members about the use of violence in the context of conflicts over national liberation and self-determination.

These divergences have made it impossible for the United Nations to conclude a Comprehensive Convention on International Terrorism that incorporates a single, all-encompassing, legally binding, criminal law definition of terrorism. The international community has adopted a series of sectorial conventions that define and criminalize various types of terrorist activities.

Since 1994, the United Nations General Assembly has repeatedly condemned terrorist acts using the following political description of terrorism:

> Criminal acts intended or calculated to provoke a state of terror in the public, a group of persons or particular persons for political purposes are in any circumstance unjustifiable, whatever the considerations of a political, philosophical, ideological, racial, ethnic, religious or any other nature that may be invoked to justify them.
>
> **United Nations**
> *50/53. Measures to Eliminate International Terrorism*

U.S. Code Title 22 Chapter 38, Section 2656f (d), defines terrorism as "Premeditated, politically motivated violence perpetrated against noncombatant targets by subnational groups or clandestine agents, usually intended to influence an audience."

Bruce Hoffman (2006), a scholar, has noted:

> It is not only individual agencies within the same governmental apparatus that cannot agree on a single definition of terrorism. Experts and other long-established scholars in the field are equally incapable of reaching a consensus.
>
> **Bruce Hoffman**
> *Inside Terrorism*

In the first edition of his magisterial survey, "Political Terrorism: A Research Guide," Alex Schmid (1983) devoted more than a hundred pages to examining more than a hundred different definitions of terrorism in an effort to discover a broadly acceptable, reasonably comprehensive explication of the word. Four years and a second edition later, Schmid was no closer to the goal of his quest, conceding in the first sentence of the revised volume that the "search for an adequate definition is still on." Walter Laqueur (2001) despaired of defining terrorism in both editions of his monumental work on the subject, maintaining that it is neither possible to do so nor worthwhile to make the attempt.

Each act of terrorism is a "performance" devised to affect many large audiences. Terrorists also attack national symbols, to show power and to attempt to shake the foundation of the country or society they are opposed to. This may negatively affect a government, while increasing the prestige of the given terrorist group and/or ideology behind a terrorist act.

Terrorist acts frequently have a political purpose. This is often where the interrelationship between terrorism and religion occurs.

When a political struggle is integrated into the framework of a religious or "cosmic" struggle, such as over the control of an ancestral homeland or holy site such as Israel and Jerusalem, failing in the political goal (nationalism) becomes equated with spiritual failure, which, for the highly committed, is worse than their own death or the deaths of innocent civilians.

Their suffering accomplishes the terrorists' goals of instilling fear, getting their message out to an audience, or otherwise satisfying the demands of their often radical religious and political agendas.

Some official, governmental definitions of terrorism use the criterion of the illegitimacy or unlawfulness of the act to distinguish between actions authorized by a government (and thus "lawful") and those of other actors, including individuals and small groups. For example, carrying out a strategic bombing on an enemy city, which is designed to affect civilian support for a cause, would not be considered terrorism if it were authorized by a government. Criterion is inherently problematic and is not universally accepted, because it denies the existence of state terrorism; the same act may or may not be classed as terrorism depending on whether its sponsorship is traced to a "legitimate" government; "legitimacy" and "lawfulness" are subjective, depending on the perspective of one government or another; and it diverges from the historically accepted meaning and origin of the term.

According to Ali Khan, the distinction lies ultimately in a political judgment (Mughal, 2012).

An associated, and arguably more easily definable, but not equivalent term is violent nonstate actor. The semantic scope of this term includes not only "terrorists," but while excluding some individuals or groups who have previously been described as "terrorists" also explicitly excludes state terrorism.

Barack Obama, commenting on the Boston Marathon bombings of April 2013, declared that "Anytime bombs are used to target innocent civilians, it is an act of terror." Various commentators have pointed out the distinction between "act of terror" and "terrorism," particularly when used by the White House. 18 U.S.C. § 2331 defines "international terrorism" and "domestic terrorism" for purposes of Chapter 113B of the Code, entitled "Terrorism":

"International terrorism" means activities with the following three characteristics:

Involve violent acts or acts dangerous to human life that violate federal or state law; Appear to be intended (i) to intimidate or coerce a civilian population; (ii) to influence the policy of a government by intimidation or coercion; or (iii) to affect the conduct of a government by mass destruction, assassination, or kidnapping; and Occur primarily outside the territorial jurisdiction of the U.S., or transcend national boundaries in terms of the means by which they are accomplished, the persons they appear intended to intimidate or coerce, or the locale in which their perpetrators operate or seek asylum.

By distinguishing terrorists from other types of criminals and terrorism from other forms of crime, we come to appreciate that terrorism is:

- Ineluctably political in aims and motives;
- Violent—or, equally important, threatens violence;
- Designed to have far-reaching psychological repercussions beyond the immediate victim or target;
- Conducted either by an organization with an identifiable chain of command or conspiratorial cell structure (whose members wear no uniform or identifying insignia) or by individuals or a small collection of individuals directly influenced, motivated, or inspired by the ideological aims or example of some existent terrorist movement and/or its leaders; and by a subnational group or non-state entity.

A definition proposed by Carsten Bockstette (2008) at the George C. Marshall European Center for Security Studies underlines the psychological and tactical aspects of terrorism:

> Terrorism is defined as political violence in an asymmetrical conflict that is designed to induce terror and psychic fear (sometimes indiscriminate) through the violent victimization and destruction of noncombatant targets (sometimes iconic symbols). Such acts are meant to send a message from an illicit clandestine organization. The purpose of terrorism is to exploit the media in order to achieve maximum attainable publicity as an amplifying force multiplier in order to influence the targeted audience(s) in order to reach short- and midterm political goals and/or desired long-term end states.
>
> **Carsten Bockstette**
> *Jihadist Terrorist Use of Strategic Communication Management Techniques*

PEJORATIVE USE

The terms "terrorism" and "terrorist" (someone who engages in terrorism) carry strong negative connotations. These terms are often used as political labels; to condemn violence or the threat of violence by certain actors as immoral, indiscriminate, and unjustified; or to condemn an entire segment of a population (Figures 10.1 and 10.2).

Those labeled "terrorists" by their opponents rarely identify themselves as such and typically use other terms or terms specific to their situation, such as separatist, freedom fighter, liberator, revolutionary, vigilante, militant, paramilitary, guerrilla, rebel, patriot, or any similar-meaning word in other languages and cultures. *Jihadi*,

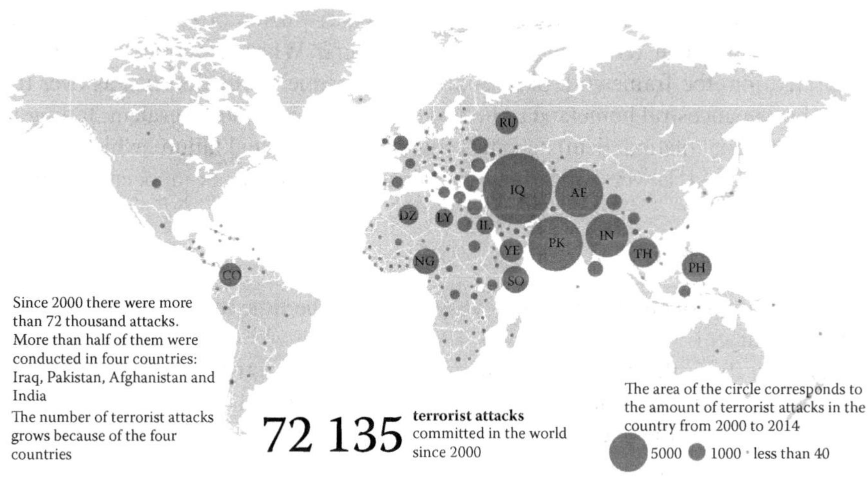

FIGURE 10.1 Terrorism incidents.

FIGURE 10.2 Terrorism acts of violence.

mujaheddin, and *fedayeen* are similar Arabic words that have entered the English lexicon. It is common for both parties in a conflict to describe each other as terrorists.

On whether particular terrorist acts, such as killing noncombatants, can be justified as the lesser evil in a particular circumstance, philosophers have expressed different views. While, according to David Rodin (2007), utilitarian philosophers can (in theory) conceive of cases in which the evil of terrorism is outweighed by the good that could not be achieved in a less morally costly way, in practice, the "harmful effects of undermining the convention of non-combatant immunity is thought to outweigh the goods that may be achieved by particular acts of terrorism."

Among the nonutilitarian philosophers, Michael Walzer (2016) argued that terrorism can be morally justified in only one specific case: when "a nation or community faces the extreme threat of complete destruction and the only way it can preserve itself is by intentionally targeting non-combatants, then it is morally entitled to do so."

Terrorist acts frequently have a political purpose. This is often where the interrelationship between terrorism and religion occurs. When a political struggle is integrated into the framework of a religious or "cosmic" struggle, such as over the control of an ancestral homeland or holy site such as Israel and Jerusalem, failing in the political goal (nationalism) becomes equated with spiritual failure, which, for the highly committed, is worse than their own death or the deaths of innocent civilians.

TYPES OF TERRORISM

Depending on the country, the political system, and the time in history, the types of terrorism vary.

In early 1975, the Law Enforcement Assistant Administration in the United States formed the National Advisory Committee on Criminal Justice Standards and Goals. One of the five volumes that the committee wrote was titled "Disorders and Terrorism," produced by the Task Force on Disorders and Terrorism under the direction of H. H. A. Cooper, Director of the Task Force staff (Cooper, 2017).

The Task Force defines terrorism as "a tactic or technique by means of which a violent act or the threat thereof is used for the prime purpose of creating overwhelming fear for coercive purposes."

It classified disorders and terrorism into six categories:

1. Civil disorder—A form of collective violence interfering with the peace, security, and normal functioning of the community.
2. Political terrorism—Violent criminal behavior designed primarily to generate fear in the community, or a substantial segment of it, for political purposes.
3. Nonpolitical terrorism—Terrorism that is not aimed at political purposes but which exhibits "conscious design to create and maintain a high degree of fear for coercive purposes, but the end is individual or collective gain rather than the achievement of a political objective."
4. Quasi-terrorism—The activities incidental to the commission of crimes of violence that are similar in form and method to genuine terrorism but which nevertheless lack its essential ingredient. It is not the main purpose of the quasi-terrorists to induce terror in the immediate victim as in the case of genuine terrorism, but the quasi-terrorist uses the modalities and techniques of the genuine terrorist and produces similar consequences and reaction. For example, the fleeing felon who takes hostages is a quasi-terrorist, whose methods are similar to those of the genuine terrorist but whose purposes are quite different.
5. Limited political terrorism—Genuine political terrorism is characterized by a revolutionary approach; limited political terrorism refers to "acts of terrorism which are committed for ideological or political motives but which are not part of a concerted campaign to capture control of the state" (http://preventwmd.org, 2017).
6. Official or state terrorism—"Referring to nations whose rule is based upon fear and oppression that reach similar to terrorism or such proportions"

(http://preventwmd.org, 2017). It may also be referred to as structural terrorism, defined broadly as terrorist acts carried out by governments in pursuit of political objectives, often as part of their foreign policy.

Other sources have defined the typology of terrorism in different ways, for example, broadly classifying it into domestic terrorism and international terrorism, or using categories such as vigilante terrorism or insurgent terrorism. One way the typology of terrorism may be defined:

- Political terrorism
- Substate terrorism
- Social revolutionary terrorism
- Nationalist–separatist terrorism
- Religious extremist terrorism
  - Religious fundamentalist terrorism
  - New religions terrorism
- Right-wing terrorism
- Left-wing terrorism
- State-sponsored terrorism
- Regime or state terrorism
- Criminal terrorism
- Pathological terrorism

MOTIVATIONS OF TERRORISTS

Attacks on "collaborators" are used to intimidate people from cooperating with the state in order to undermine state control. This strategy was used in Ireland, in Kenya, in Algeria, and in Cyprus during their independence struggles.

Attacks on high-profile symbolic targets are used to incite counterterrorism by the state to polarize the population. This strategy was used by Al-Qaeda in its attacks on the World Trade Center and the Pentagon in the United States on September 11, 2001. These attacks are also used to draw international attention to struggles that are otherwise unreported, such as the Palestinian airplane hijackings in 1970 and the South Moluccan hostage crisis in the Netherlands in 1975.

Terrorist organizations do not select terrorism for its political effectiveness. Individual terrorists tend to be motivated more by a desire for social solidarity with other members of their organization than by political platforms or strategic objectives, which are often murky and undefined. Additionally, possible relationships between the type of economy within a country and ideology are associated with terrorism. Some terrorists like Timothy McVeigh were motivated by revenge against a state for its actions against its citizens.

Democracy and Domestic Terrorism

The relationship between domestic terrorism and democracy is very complex. Terrorism is most common in nations with intermediate political freedom, and it is least common in the most democratic nations. However, one study suggests that suicide terrorism may be an exception to this general rule. Evidence regarding this particular method of terrorism reveals that every modern suicide campaign has targeted a democracy—a state with a considerable degree of political freedom. The study suggests that concessions awarded to terrorists during the 1980s and 1990s for suicide attacks increased their frequency (Hamdan, 2015). There is a connection between the existence of civil liberties, democratic participation, and terrorism. According to Young and Dugan, these things encourage terrorist groups to organize and generate terror.

Some examples of terrorism in nondemocratic nations include ETA in Spain under Francisco Franco (although the group's terrorist activities increased sharply after Franco's death), the Organization of Ukrainian Nationalists in prewar Poland, the Shining Path in Peru under Alberto Fujimori, the Kurdistan Workers Party when Turkey was ruled by military leaders, and the ANC in South Africa. Democracies, such as Japan, the United Kingdom, the United States, Israel, Indonesia, India, Spain, Germany, and the Philippines, have also experienced domestic terrorism.

While a democratic nation espousing civil liberties may claim a sense of higher moral ground than other regimes, an act of terrorism within such a state may cause a dilemma: whether to maintain its civil liberties and thus risk being perceived as ineffective in dealing with the problem or, alternatively, to restrict its civil liberties and thus risk delegitimizing its claim of supporting civil liberties. For this reason, homegrown terrorism has started to be seen as a greater threat, as stated by former Central Intelligence Agency Director Michael Hayden (2015). This dilemma, some social theorists would conclude, may very well play into the initial plans of the acting terrorist(s), namely, to delegitimize the state and cause a systematic shift toward anarchy via the accumulation of negative sentiments toward the state system.

Religious Terrorism

Religious terrorism is terrorism performed by groups or individuals, the motivation of which is typically rooted in faith-based tenets. Terrorist acts throughout history have been performed on religious grounds with the goal to either spread or enforce a system of belief, viewpoint, or opinion. The validity and scope of religious terrorism are limited to the individual or a group view or interpretation of that belief system's teachings.

Intimate Terrorism

Intimate terrorism (IT), also known as domestic abuse, may also involve emotional and psychological abuse. IT is one element in a general pattern of control by one partner over the other. IT is more likely to escalate over time, not as likely to be mutual, and more likely to involve serious injury. IT batterers include two types: "generally violent-antisocial" and "dysphoric–borderline." The first type includes people with general psychopathic and violent tendencies. The second type includes people who are emotionally dependent on the relationship. Violence by a person against his or her intimate partner is often done as a way to control his or her partner, even if this kind of violence is not the most frequent.

PERPETRATORS

The perpetrators of acts of terrorism can be individuals, groups, or states. According to some definitions, clandestine or semiclandestine state actors may also carry out terrorist acts outside the framework of a state of war. However, the most common image of terrorism is that it is carried out by small and secretive cells, highly motivated to serve a particular cause, and many of the most deadly operations in recent times, such as the September 11 attacks, the London underground bombing, the 2008 Mumbai attacks, and the 2002 Bali bombing were planned and carried out by a close clique, composed of close friends, family members, and other strong social networks. These groups benefited from the free flow of information and efficient telecommunications to succeed where others had failed.

To avoid detection, a terrorist will look, dress, and behave normally until executing the assigned mission. Some claim that attempts to profile terrorists based on personality, physical, or sociological traits are not useful. The physical and behavioral description of the terrorist could describe almost any normal person. However, the majority of terrorist attacks are carried out by military age men, aged 16–40 (martinslibrary.blogspot.com/2014/08/terrorism-types-of-terrorism).

Nonstate Groups

Groups not part of the state apparatus or in opposition to the state are most commonly referred to as a "terrorist" in the media.

State Sponsors

A state can sponsor terrorism by funding or harboring a terrorist group. Opinions as to which acts of violence by states consist of state-sponsored terrorism vary widely. When states provide funding for groups considered by some to be terrorist, they rarely acknowledge them as such.

STATE TERRORISM

Civilization is based on a clearly defined and widely accepted yet often unarticulated hierarchy. Violence done by those higher on the hierarchy to those lower is nearly always invisible, that is, unnoticed. When it is noticed, it is fully rationalized. Violence done by those lower on the hierarchy to those higher is unthinkable and, when it does occur, is regarded with shock, horror, and the fetishization of the victims.

State terrorism has been used to refer to terrorist acts committed by governmental agents or forces. This involves the use of state resources employed by a state's foreign policies, such as using its military to directly perform acts of terrorism, examples of which include the German bombing of London, the Japanese bombing of Pearl Harbor, the British firebombing of Dresden, and the U.S. atomic bombing of Hiroshima during World War II. The use of terror tactics is common in international relations, and the state has been and remains a more likely employer of terrorism

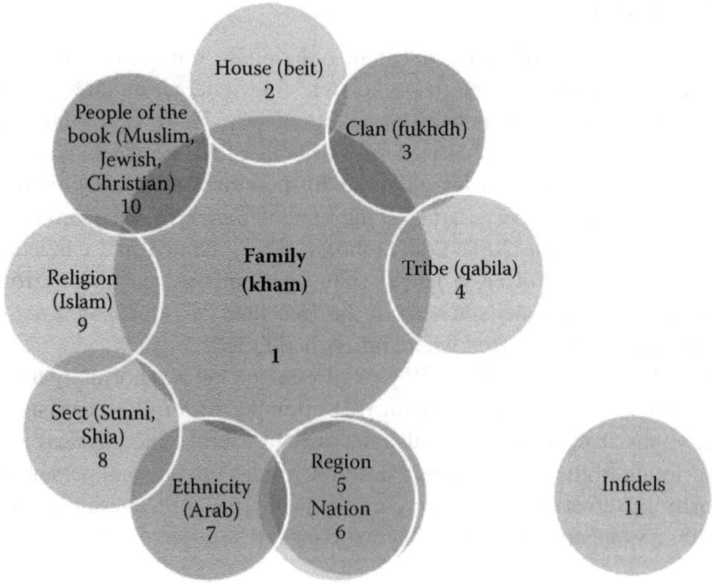

FIGURE 10.3 Muslim hierarchy of allegiance.

within the international system than insurgents. The first strike option as an example of the "terror of coercive diplomacy" as a form of this, which holds the world hostage with the implied threat of using nuclear weapons in "crisis management" and argues that the institutionalized form of terrorism has occurred as a result of changes that took place following World War II.

In this analysis, state terrorism exhibited as a form of foreign policy was shaped by the presence and use of weapons of mass destruction and that the legitimizing of such violent behavior led to an increasingly accepted form of this state behavior.

The concept is also used to describe political repressions by governments against their own civilian populations with the purpose of inciting fear. For example, taking and executing civilian hostages or extrajudicial elimination campaigns are commonly considered "terror" or terrorism, for example, during the Red Terror or the Great Terror. Such actions are also often described as democide or genocide, which have been argued to be equivalent to state terrorism. Empirical studies on this have found that democracies have little democide.

Understanding the structure of the Muslim hierarchy of allegiance will help one understand the terrorist cell structure, their duty, and the flow of information and obedience (Figure 10.3) (Ansarian, 2010).

TACTICS

Terrorism is a form of asymmetric warfare and is more common when direct conventional warfare will not be effective because forces vary greatly in power.

The context in which terrorist tactics are used is often a large-scale, unresolved political conflict. The type of conflict varies widely; historical examples include the following:

- Secession of a territory to form a new sovereign state or become part of a different state
- Dominance of territory or resources by various ethnic groups
- Imposition of a particular form of government
- Economic deprivation of a population
- Opposition to a domestic government or occupying army
- Religious fanaticism

Terrorist attacks are often targeted to maximize fear and publicity, usually using explosives or poison. There is concern about terrorist attacks employing weapons of mass destruction. Terrorist groups usually methodically plan attacks in advance and may train participants, plant undercover agents, and raise money from supporters or through organized crime. Communications occur through modern telecommunications or through old-fashioned methods such as couriers.

RESPONSES

One posting a sign notifying shoppers with the increase of surveillance is due to a perceived increase in the risk of a terrorism threat.

Responses to terrorism are broad in scope. They can include realignments of the political spectrum and reassessments of fundamental values.

Specific types of responses include the following:

- Targeted laws, criminal procedures, deportations, and enhanced police powers
- Target hardening, such as locking doors or adding traffic barriers
- Preemptive or reactive military action
- Increased intelligence and surveillance activities
- Preemptive humanitarian activities
- More permissive interrogation and detention policies

The term "counterterrorism" has a narrower connotation, implying that it is directed at terrorist actors (Global Terrorism Database, http://www.start.umd.edu/gtd).

DATABASES

The following terrorism databases are or were made publicly available for research purposes and track specific acts of terrorism (Global Terrorism Database, http://www.start.umd.edu/gtd):

- Global Terrorism Database, an open-source database on terrorist events around the world from 1970 through 2015, with more than 150,000 cases
- MIPT Terrorism Knowledge Base

- Worldwide Incidents Tracking System
- Tocsearch (dynamic database)

The following is a publicly available resource index electronic and bibliographic resource on the subject of terrorism:

- Human Security Gateway

The following terrorism databases are maintained in secrecy by the U.S. Government for intelligence and counterterrorism purposes:

- Terrorist Identities Datamart Environment
- Terrorist Screening Database

WAR ON TERROR

The War on Terror (WoT), also known as the Global War on Terrorism, is a metaphor of war referring to the international military campaign that started after the September 11 attacks on the United States. U.S. President George W. Bush first used the term "War on Terror" on September 20, 2001. The Bush administration and the western media have since used the term to argue a global military, political, legal, and conceptual struggle against both organizations designated terrorist and regimes accused of supporting them. It was originally used with a particular focus on countries associated with Islamic terrorism organizations including al-Qaeda and like-minded organizations.

In 2013, President Barack Obama announced that the United States was no longer pursuing a WoT, as the military focus should be on specific enemies rather than a tactic. He stated, "We must define our effort not as a boundless 'Global War on Terror,' but rather as a series of persistent, targeted efforts to dismantle specific networks of violent extremists that threaten America."

U.S. Objectives

- NATO
- Trans Sahara initiative
- Major military operations (Afghanistan • Pakistan • Iraq • Somalia • Yemen)
- Allies involved in major operations

The Authorization for Use of Military Force against Terrorists, or AUMF, was made law on September 14, 2001, to authorize the use of U.S. Armed Forces against those responsible for the attacks on September 11, 2001. It authorized the President to use all necessary and appropriate force against those nations, organizations, or persons he determines planned, authorized, committed, or aided the terrorist attacks that occurred on September 11, 2001, or harbored such organizations or persons, in order to prevent any future acts of international terrorism against the United States by such nations, organizations, or persons. The congress declares this is intended to constitute specific statutory authorization within the meaning of section 5(b) of the War Powers Resolution of 1973.

The George W. Bush administration defined the following objectives in the WoT:

- Defeat terrorists such as Osama bin Laden, Abu Musab al-Zarqawi, and demolish their organizations
- Identify, locate, and demolish terrorists along with their organizations
- Deny sponsorship, support, and sanctuary to terrorists
- End the state sponsorship of terrorism
- Establish and maintain an international standard of accountability with regard to combating terrorism
- Strengthen and sustain the international effort to combat terrorism
- Work with willing and able states
- Enable weak states
- Persuade reluctant states
- Compel unwilling states
- Interdict and disorder material support for terrorists
- Abolish terrorist sanctuaries and havens
- Diminish the underlying conditions that terrorists seek to exploit
- Partner with the international community to strengthen weak states and prevent (re)emergence of terrorism
- Win the war of ideals
- Defend U.S. citizens and interests at home and abroad
- Integrate the National Strategy for Homeland Security
- Attain domain awareness
- Enhance measures to ensure the integrity, reliability, and availability of critical, physical, and information-based infrastructures at home and abroad
- Implement measures to protect U.S. citizens abroad
- Ensure an integrated incident management capability

Terrorism and counterterrorism research is an interdisciplinary academic field that seeks to understand the causes of terrorism, how to prevent it as well as its impact in the broadest sense. Terrorism research can be carried out in both military and civilian contexts.

11 Managing Terrorism Threat/Vulnerability Assessments and Risk Analysis

With the increased threat of terrorism, public, private, and governmental agencies face an increased need to understand and manage the risk to their employees and organizational assets. A three-tiered terrorism risk management plan, which includes initial and detailed assessments and a variety of risk management techniques, can effectively reduce the terrorism risk.

Over the past decade, the changing political landscape has dictated that public, private, and governmental organizations not only understand terrorism risk but also develop a proactive plan to assess and/or manage this risk. As demonstrated by the attack on the Alfred P. Murrah Federal Building in Oklahoma City in 1995, simply being located in a less populous city or region does not guarantee safety. In fact, it is the lower profile targets with lower levels of security awareness that may appear to be the "softer" target to a potential terrorist.

In addition to the terrible loss of life, the events in 2001 also highlighted the financial costs of terrorism. As with other natural and manmade hazards, financial coverage may not be available and/or affordable to protect the financial assets of an organization when faced with a terrorism threat.

DEVELOPING A TERRORISM RISK MANAGEMENT PROGRAM

Managing risk associated with the threat of terrorism can be a daunting task for companies. Some of the most common questions include how to begin a terrorism risk management program, what assets should be protected, and what are the most effective mitigation solutions.

Similar to managing the risk from other hazards, a terrorism risk management program should provide a logical and systematic framework for identifying and dealing with potential terrorist threats. A three-phase terrorist risk management program, detailed below, can be used as a framework for establishing a terrorism risk management program.

Phase I—Threat Identification and Initial Site Assessment

Understanding the type, source, and probability associated with different threats is an important element in the program. The key elements of the threat identification phase include the following.

1. Threat recognition and identification
2. Threat potential
3. Site security assessment

While many organizations have some knowledge of the various threats facing their facilities and employees, many do not recognize how vulnerable their sites actually are until a detailed assessment is performed.

One of the products that may result from the site security assessment is a detailed site survey that includes the determination of the effective standoff distance at all exposed sides of the building perimeter. Even for sites that are considered "secure," existing security measures are often found to be insufficient to deter a well-planned terrorist attack.

Phase II—Detailed Risk Assessment

The information gathered in the Phase I assessment can then be used to focus organizational resources to determine the impact of a particular terrorist event on the facility. Analyses that may form part of the detailed risk assessment include the following.

- Blast and explosion analysis
- Progressive collapse (structural stability) analysis
- Chemical, biological, and radiological (CBR) threat assessment

The blast, progressive collapse, and chemical/biological analyses can provide a detailed assessment of the threat to the structures, nonstructural elements, and the employees. Examples of these analyses are shown in Tables 11.1 and 11.2. Software tools are available to estimate the impact and dispersion of toxic releases

TABLE 11.1
Probability Levels of an Undesired Event

Probability Level	Specific Event
A: Frequent	Likely to occur frequently
B: Probable	Will occur several times
C: Occasional	Likely to occur sometimes
D: Remote	Unlikely but possible to occur
E: Improbable	So unlikely it can be assumed occurrence may not be experienced

TABLE 11.2
Severity Levels of Undesired Event Consequences

Severity Level	Characteristics
I. Catastrophic	Death, system loss, or severe environmental damage
II. Critical	Severe injury, severe occupational illness, major system or environmental damage
III. Marginal	Minor injury, minor occupational illness, or minor system or environmental damage
IV. Negligible	Less than minor injury, occupational illness, or less than minor system or environmental damage

and produce similar types of contours as a result of a biological or chemical terrorist attack.

Phase III—Risk Management

Once the risks have been identified and assessed, putting a comprehensive risk management plan in place for terrorism risk is similar in many respects to understanding and managing the risks due to other hazards, such as extreme wind or earthquakes. Often, emergency planning and disaster recovery preparations that are in place for other types of hazards can be extended to prepare for and/or protect against terrorist attacks.

A comprehensive terrorism risk management plan should, at a minimum, include the following components.

- Protection of the facility and its occupants
- Emergency planning and disaster recovery
- Reduction of financial risk

Protection of the building occupants through the implementation of physical or electronic security measures, the application of window film to reduce glazing hazards, and increased employee awareness is often a first step in many terrorism risk management plans. Reducing the financial risk associated with a terrorism event may be a more challenging issue. As with other natural and manmade hazards, the cost of insurance for losses associated with a terrorism event, if available, may have risen to a level that is no longer affordable. The financial exposure may need to be addressed through a combination of risk mitigation measures, alternate or back-up facilities, and insurance.

Threat and risk assessments are widely recognized as valid decision support tools to establish and prioritize security program requirements.

A threat analysis, the first step in determining risk, identifies and evaluates each threat on the basis of various factors, such as its capability and intent to attack an asset, the likelihood of a successful attack, and its lethality.

Risk management is the deliberate process of understanding "risk"—the likelihood that a threat will harm an asset with some severity of consequences—and

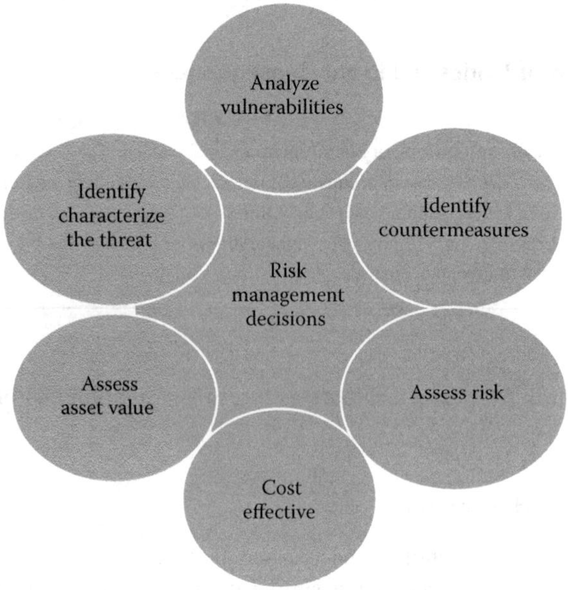

FIGURE 11.1 Factors considered in making risk management decisions.

deciding on and implementing actions to reduce it (Figure 11.1). Risk management principles acknowledge the following:

1. While risk generally cannot be eliminated, it can be reduced by enhancing protection from validated and credible threats.
2. Although many threats are possible, some are more likely to occur than others.
3. All assets are not equally critical.

Generally, the risk assessment process is a deliberate, analytical approach to identify which threats can exploit which vulnerabilities in an organization's specific assets.

These variables are ranked according to predetermined criteria, such as the probability of a threat targeting a specific asset or the impact of a vulnerability being exploited by a specific threat. The risk assessment results in a prioritized list of risks (i.e., threat–asset–vulnerability combinations) that can be used to select safeguards to reduce vulnerabilities and create a certain level of protection.

The following are challenges to applying an accepted threat and risk assessment process to cities and infrastructure:

1. Security issues (e.g., revealing intelligence sources and methods) related to providing valid threat data from the intelligence community to city officials
2. The lack of specificity in the intelligence community's threat information

3. The complexity and magnitude of a large city as a subject of a threat and risk assessment. These challenges could be overcome through federal-city collaboration.

The risk assessment generates specific threat scenarios from valid intelligence and threat data and pairs them with vulnerabilities in its critical assets. The multidisciplinary risk assessment then assigns weights or values to these threat–asset vulnerability pairings according to the likelihood of such events occurring and the consequences of assets being compromised or attacked. This process is based on a Department of Defense military standard and work by the Department of Transportation's Volpe National Transportation Systems Center.

The Federal Bureau of Investigation (FBI) is in the best position to take the federal lead in facilitating city-specific threat and risk assessments. The FBI, through the Attorney General, is the lead agency for domestic terrorism crisis management. As a member of the intelligence community, the FBI also collects, analyzes, and reports threat information on domestic origin threats and targets.

Cities are larger and more complex than most entities subject to threat and risk assessments, such as military bases, ports, and petroleum processing facilities. However, size and complexity would not preclude conducting threat and risk assessments.

Threat and risk assessments are widely recognized as effective decision support tools for prioritizing security investments, and we identified several public and private sector organizations that use them.

While it is not possible to reduce risk to all potential targets against weapon of mass destruction (WMD) terrorism, risk assessments can help ensure that training, equipment, and other safeguards are justified and implemented based on threat, the vulnerability of the asset to an attack, and the importance of the asset.

A vulnerability assessment should consist of experienced military and civilian personnel from a range of disciplines, including structural and civil engineering, security, and operations readiness. The assessment team uses a risk assessment model to identify and rank order a site's strengths and weaknesses. The specific elements of the model include asset criticality, site vulnerability, and the ease with which a threat can gain access to an asset.

A comprehensive vulnerability assessment of the U.S. surface transportation and inner infrastructure (i.e., road, rail, transit, pipeline, and maritime) was conducted (Transportation Security Administration, 2016). The goal of the assessment is to

1. Identify and rank key threats to and critical vulnerabilities of the national transportation infrastructure
2. Recommend possible countermeasures to improve infrastructure protection from a host of threats such as terrorism, accidents, and natural disasters

Step 1: Identify Threats and Pair with Assets

Threat identification is the most important step in the risk assessment process. If threats are not accurately identified, the risks they represent cannot be reduced or eliminated. Threats that the company is concerned with include trusted or incompetent insiders, criminals, terrorists, and environmental and system-induced threats. In characterizing the threat, the infrastructure examines the historical record of security and safety breaches and obtains location-specific threat information from the intelligence community and open sources. These threats are then paired with company assets that represent likely targets.

Step 2: Identify Asset Vulnerabilities

The risk assessment team identifies weaknesses in the infrastructures critical assets that could be exploited by the threats identified in Step 1 and determines their nature and source. Methods used to identify vulnerabilities include evaluating data obtained through surveys and historical data from related incidents and applying formal vulnerability analysis techniques. Asset vulnerabilities can include operations and processes, policies and procedures, physical and technical security, information security, personnel security, and operations security.

Step 3: Determine Risk Through Scenarios

The risk assessment team develops credible risk scenarios to describe how undesired events may occur and to determine the effect of each undesired event on the infrastructure assets. The set of scenarios may not be an exhaustive list of all possible undesired events, but each valid threat that has been identified should be represented in at least one scenario.

Step 4: Identify Actions, as Necessary, That Lead to Risk Reduction

Countermeasures are actions that either eliminate the causes or reduce the effects of one or more vulnerabilities. Countermeasures could include additional checkpoints controlling access to a facility, security cameras, personnel background investigations, new procedures, or chemical protective gear. Countermeasures are identified and inserted into a scenario, and the risk rating for that scenario is recalculated to account for the effect of the countermeasure.

The countermeasures are selected on the basis of factors such as whether they reduce the probability of an undesired event occurring, their implementing cost, and any additional enforcement and audit requirements. Countermeasures can be prioritized by considering a number of factors, including the amount of resulting risk reduction, cost, difficulty to implement, or a combination thereof. Usually, there is a point beyond which adding countermeasures will raise costs without appreciably enhancing the protection afforded.

In closing as part of that effort on exploring how some public and private sector organizations can establish the requirements and prioritize those allocated resources to support the incident and safeguard those assets against a variety of threats through prior planning and procedures design to lessen the impact on those public and private sector organizations to include the threat of terrorism.

1. Examined threat and risk assessment approaches used by several public and private sector organizations to deal with terrorist and other security risks and obtained detailed information on a private company's risk-assessment process.

2. Determined whether training and assistance used threat and risk assessments to establish requirements for dealing with WMD terrorist incidents.
3. Assessed the challenges of using formal threat and risk assessments to help define requirements and prioritize, and target resources.

INFRASTRUCTURE FACILITY DESIGNED USING THE CRIME PREVENTION THROUGH ENVIRONMENTAL DESIGN PROCESS

The risk management plan should have a capability for determining changes in risk due to threat information or changes to security operations and building protection. The regimented procedure will maintain a focus on effectiveness and prevent fragmented decision making for risk reduction.

With the increased threat of terrorism, both in the United States and abroad, public, private, and governmental agencies face an increased need to understand and manage the risk to their employees and organizational assets due to the terrorism risk. A three-tiered terrorism risk management plan, which includes initial and detailed assessments and a variety of risk management techniques, can be implemented to effectively reduce the risk from a terrorist attack.

THREAT/VULNERABILITY ASSESSMENTS AND RISK ANALYSIS

All facilities face a certain level of risk associated with various threats. These threats may be the result of natural events, accidents, or intentional acts to cause harm. Regardless of the nature of the threat, facility owners have a responsibility to limit or manage risks from these threats to the extent possible.

The federal government has implemented The Risk Management Process for Federal Facilities: An Interagency Security Committee Standard (Integrated Security Committee, 2016), which states,

> Risk is a function of the values of threat, consequence, and vulnerability. The objective of risk management is to create a level of protection that mitigates vulnerabilities to threats and the potential consequences, thereby reducing risk to an acceptable level. A variety of mathematical models are available to calculate risk and to illustrate the impact of increasing protective measures on the risk equation.

Facility owners, particularly owners of public facilities, should develop and implement a security risk management methodology that adheres to the Interagency Security Committee (ISC) standard while also supporting the security needs of the organization. Landlords who desire to lease space to federal government agencies should implement the ISC standard in the design of new facilities and/or the renovation of existing facilities.

THREAT ASSESSMENT

The first step in a risk management program is a threat assessment. A threat assessment considers the full spectrum of threats (i.e., natural, criminal, terrorist, accidental, etc.) for a given facility/location. The ISC standard addresses only manmade threats, but individual agencies are free to expand upon the threats they consider. The assessment should examine supporting information to evaluate the relative likelihood of occurrence for each threat. For natural threats, historical data concerning frequency of occurrence for given natural disasters such as tornadoes, hurricanes, floods, fire, or earthquakes can be used to determine the credibility of the given threat.

For criminal threats, the crime rates in the surrounding area provide a good indicator of the type of criminal activity that may threaten the facility. In addition, the type of assets and/or activity located in the facility may also increase the target attractiveness in the eyes of the aggressor. The type of assets and/or activity located in the facility will also relate directly to the likelihood of various types of accidents. For example, a facility that utilizes heavy industrial machinery will be at higher risk for serious or life-threatening job related accidents than a typical office building.

For terrorist threats, the attractiveness of the facility as a target is a primary consideration. In addition, the type of terrorist act may vary based on the potential adversary and the method of attack most likely to be successful for a given scenario. For example, a terrorist wishing to strike against the federal government may be more likely to attack a large federal building than to attack a multitenant office building containing a large number of commercial tenants and a few government tenants.

However, if security at the large federal building makes mounting a successful attack too difficult, the terrorist may be diverted to a nearby facility that may not be as attractive from an occupancy perspective but has a higher probability of success due to the absence of adequate security. In general, the likelihood of terrorist attacks cannot be quantified statistically since terrorism is, by its very nature, random. Specific definitions are important to quantify the level of each threat.

The more specific the definition, the more consistent the assessments will be especially if the assessments are being performed by a large number of assessors. Example assessments are provided below:

- *Defined:* Manmade: There are aggressors who utilize this tactic who are known to be targeting this facility or the organization. There is a history of this type of activity in the area and this facility is a known target. Specific threats have been received or identified by law enforcement agencies. Natural: Events of this nature occur in the immediate vicinity on a frequent basis.
- *Credible:* Manmade: There are aggressors who utilize this tactic who are known to target this type of facility. There is a history of this type of activity in the area, and this facility and/or similar facilities have been targets previously. No specific threat has been received or identified by law enforcement agencies. Natural: Events of this nature occur in the immediate vicinity periodically (i.e., once every 10 years).
- *Potential:* Manmade: There are aggressors who utilize this tactic, but they are not known to target this type of facility. There is a history of this type of activity in the area, but this facility has not been a target. Natural: Events of this nature occur in the region on a sporadic basis.

- *Minimal:* Manmade: No aggressors who utilize this tactic are identified for this facility and there is no history of this type of activity at the facility or the neighboring area. Natural: There is no history of this type of event in the area.

Vulnerability Assessment

Once the plausible threats are identified, a vulnerability assessment must be performed. The vulnerability assessment considers the potential impact of loss from a successful attack as well as the vulnerability of the facility/location to an attack. Impact of loss is the degree to which the mission of the agency is impaired by a successful attack from the given threat. A key component of the vulnerability assessment is properly defining the ratings for impact of loss and vulnerability.

These definitions may vary greatly from facility to facility. For example, the amount of time that mission capability is impaired is an important part of impact of loss. If the facility being assessed is an Air Route Traffic Control Tower, a downtime of a few minutes may be a serious impact of loss, while for a Social Security office, a downtime of a few minutes would be minor. A sample set of definitions for impact of loss is provided below. These definitions are for an organization that generates revenue by serving the public.

- *Devastating:* The facility is damaged/contaminated beyond habitable use. Most items/assets are lost, destroyed, or damaged beyond repair/restoration. The number of visitors to other facilities in the organization may be reduced by up to 75% for a limited period of time.
- *Severe:* The facility is partially damaged/contaminated. Examples include partial structure breach resulting in weather/water, smoke, impact, or fire damage to some areas. Some items/assets in the facility are damaged beyond repair, but the facility remains mostly intact. The entire facility may be closed for a period of up to two weeks and a portion of the facility may be closed for an extended period of time (more than one month). Some assets may need to be moved to remote locations to protect them from environmental damage. The number of visitors to this and other facilities in the organization may be reduced by up to 50% for a limited period of time.
- *Noticeable:* The facility is temporarily closed or unable to operate but can continue without an interruption of more than one day. A limited number of assets may be damaged, but the majority of the facility is not affected. The number of visitors to this and other facilities in the organization may be reduced by up to 25% for a limited period of time.
- *Minor:* The facility experiences no significant impact on operations (downtime is less than four hours) and there is no loss of major assets.

Vulnerability is defined to be a combination of the attractiveness of a facility as a target and the level of deterrence and/or defense provided by the existing countermeasures. Target attractiveness is a measure of the asset or facility in the eyes of an aggressor and is influenced by the function and/or symbolic importance of the facility. Sample definitions for vulnerability ratings are as follows:

- *Very high:* This is a high-profile facility that provides a very attractive target for potential adversaries, and the level of deterrence and/or defense provided by the existing countermeasures is inadequate.
- *High:* This is a high-profile regional facility or a moderate profile national facility that provides an attractive target and/or the level of deterrence and/or defense provided by the existing countermeasures is inadequate.
- *Moderate:* This is a moderate-profile facility (not well known outside the local area or region) that provides a potential target and/or the level of deterrence and/or defense provided by the existing countermeasures is marginally adequate.
- *Low:* This is not a high-profile facility and provides a possible target and/or the level of deterrence and/or defense provided by the existing countermeasures is adequate.

The vulnerability assessment may also include detailed analysis of the potential impact of loss from an explosive, chemical or biological attack. Professionals with specific training and experience in these areas are required to perform these detailed analyses.

Risk Analysis

A combination of the impact of loss rating and the vulnerability rating can be used to evaluate the potential risk to the facility from a given threat. A sample risk matrix is depicted in Figure 11.2. In Figures 11.3 through 11.5, high risks are designated by the dark gray cells, moderate risks by the light gray cells, and low risks by the gray cells (Figure 11.6).

Upgrade Recommendations

Based on the findings from the risk analysis, the next step in the process is to identify countermeasure upgrades that will lower the various levels of risk. If an organization has minimum standard countermeasures for a given facility level which are not currently present, these countermeasures should automatically be included in the upgrade

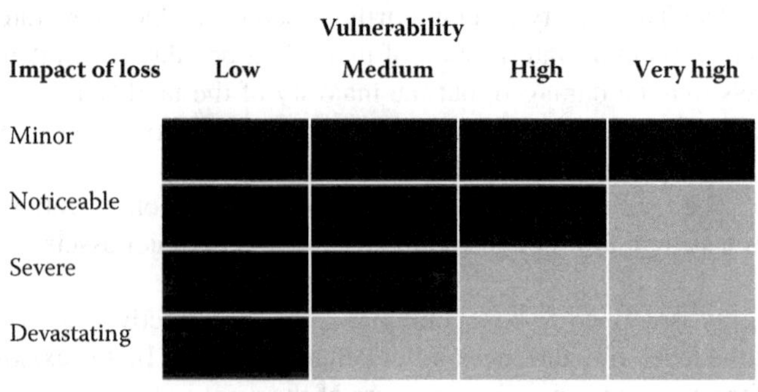

FIGURE 11.2 Matrix identifying levels of risk minimal threat.

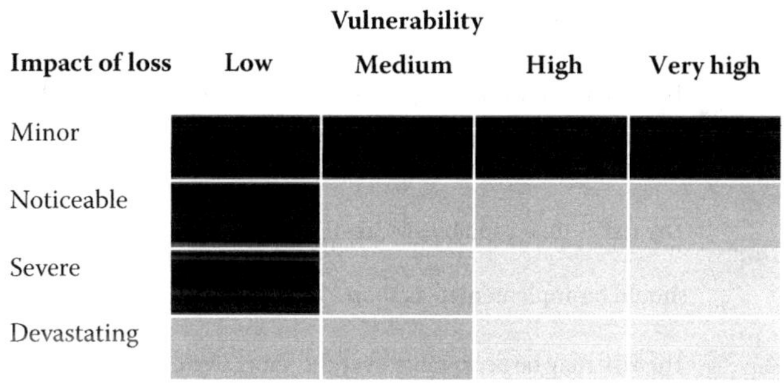

FIGURE 11.3 Potential threat.

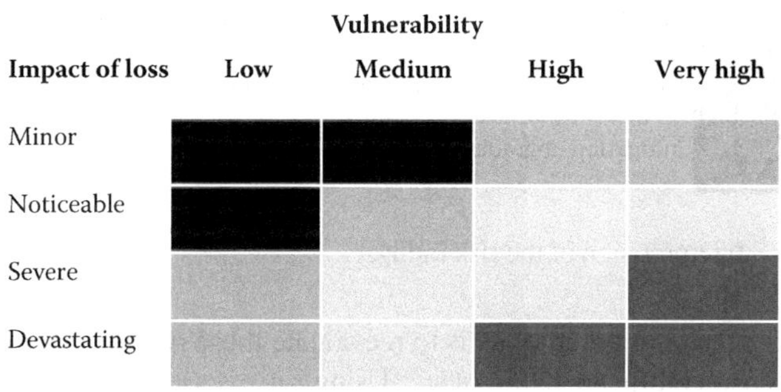

FIGURE 11.4 Credible threat.

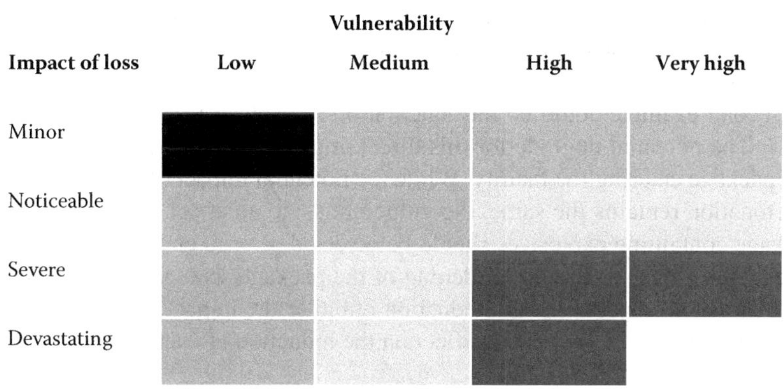

FIGURE 11.5 Defined threat.

recommendations. Additional countermeasure upgrades above the organization's recommended minimum standards should be recommended as necessary to address the specific threats and associated unacceptable risks identified for the facility.

Reevaluation of Risks

The implementation of the recommended security and/or structural upgrades should have a positive effect on the impact of loss and/or the vulnerability ratings for each

Rating category	Description
Very high	The risk is totally unacceptable. Immediate measures must be taken to reduce these risks and mitigate hazards.
High	The risk is unacceptable. Measures to reduce risk and mitigation hazards should be implemented as soon as possible.
Medium	The risk may be acceptable over the short term. Plans to reduce risk and mitigate hazards should be included in future plans and budgets.
Low	The risks are acceptable. Measures to further reduce risk or mitigate hazards should be implemented in conjunction with other security and mitigation upgrades.

FIGURE 11.6 Interpretation of the risk ratings.

threat. The final step in the process is to reevaluate these two ratings for each threat in light of the recommended upgrades. Using an exterior explosive threat as an example, the installation of window retrofits (i.e., security window film, laminated glass, etc.) will not prevent the explosive attack from occurring, but it should reduce the impact of loss/injury caused by hazardous flying glass. Therefore, the impact of loss rating for an explosive threat would improve, but the vulnerability rating would stay the same.

A second example could be introduction of an explosive into the interior of the facility. The potential upgrade for this threat might be X-ray package screening for every package entering the facility. While the potential impact of loss from an internal detonation remains the same, the vulnerability to an attack is lessened because a package containing explosives should be detected prior to entering the facility. To further reduce risk, structural hardening of the package screening areas could also reduce potential impact of loss. Reduction of either the impact of loss rating or the vulnerability rating has a positive effect on the reduction of overall risk.

APPLICATION

Threat/vulnerability assessments and risk analysis can be applied to any facility and/or organization. The federal government has been utilizing varying types of assessments and analyses for many years. Federal Security Risk Management (FSRM) is basically the process described in this paper. FSRM is currently being used by several federal agencies as well as commercial businesses to assess their facilities.

Software is available to assist in performing threat/vulnerability assessments and risk analyses. The software tool associated with implementation of FSRM is entitled

FSR-Manager (http://www.wbdg.org/resources/threat-vulnerability-assessments-and-risk-analysis). This tool is designed to be used by security personnel and allows the user to

- Input a description of the facility, including the number of people occupying the facility, the tenants represented, the contacts made during the assessment, any information gathered from the contacts, the construction details, etc.
- Input existing countermeasures
- Input threats and rate
- Assign an impact of loss to each threat
- Input countermeasure upgrade alternatives and their associated costs

The FSR-Manager uses the above inputs to

- Calculate vulnerability to each threat based on existing countermeasures.
- Determine the risk level from each threat and classify the risk level as high, medium, or low.
- Check the existing countermeasures against a list of ISC-recommended countermeasures for the given facility security level and specific threats. The user is provided a list of potential countermeasure upgrades from which the user may choose what to recommend for implementation.
- Reevaluate the vulnerability and associated risk level for each threat based on countermeasure upgrade recommendations.
- Use all of the input information to complete a template report in Microsoft Word. (Nancy A. Renfroe, 2016)

CRIME PREVENTION THROUGH ENVIRONMENTAL DESIGN

The Crime Prevention through Environmental Design (CPTED) is a proven methodology that not only enhances the performance of these security and safety measures, but also provides aesthetics and value engineering (Fennelly, 2013).

CPTED utilizes four primary, overlapping principles:

1. Natural surveillance
2. Natural access control
3. Territoriality
4. Maintenance

Natural surveillance follows the premise that criminals do not wish to be observed; placing legitimate "eyes" on the street, such as providing window views and lighting, increases the perceived risk to offenders, reduces fear for bona fide occupants and visitors, as well as lessening reliance on camera surveillance.

Natural access control supplements physical security and operational measures with walls, fences, ravines, or even hedges to define site boundaries, to channel legitimate occupants and visitors to designated entrances and to reduce access points and escape routes.

Territoriality involves strategies to project a sense of ownership to spaces such that it becomes easier to identify intruders because they don't seem to belong. Clear differentiation between public, semipublic, and private spaces by using signage, fences, pavement treatment, art, and flowers are examples of ways to express ownership.

Maintenance is a key element to preserve lines of sights for surveillance, to retain the defensiveness of physical elements, and to project a sense of care and ownership. Together, the principles of CPTED increase the effectiveness of operational, technical, and physical safety methods, thereby lessening equipment and operating costs.

For total design efficiency and cost effectiveness, security, safety, and CPTED measures are best applied at the beginning of a project. Security programming is a useful practice to identify security design requirements necessary to satisfy stakeholder concerns.

DESIGN BASIS THREAT TACTICS

Depending on the building type, acceptable levels of risk, and decisions made based on recommendations from a comprehensive threat assessment, vulnerability assessment, and risk analysis, appropriate countermeasures should be implemented to protect people, assets, and mission.

Some types of attack and threats to consider include the following:

- Unauthorized entry/trespass (forced and covert)
- Insider threats
- Explosive threats: stationary and moving vehicle-delivered, mail bombs, package bombs
- Ballistic threats: small arms, high-powered rifles, drive-by shootings, etc.
- WMDs (CBR)
- Disruptive threats (hoaxes, false reports, malicious attempts to disrupt operations)
- Cyber and information security threats

Supervisory Control and Acquisition Data system threats are relevant as they relate to HVAC, mechanical/electrical systems control, and other utility systems that are required to operate many functions within building.

Unauthorized Entry (Forced and Covert)

Protecting the facility and assets from unauthorized persons is an important part of any security system. Some items to consider include the following:

- Compound or facility access control
  - Control perimeter: fences, bollards, antiram barriers
  - Traffic control, remote controlled gates, antiram hydraulic drop arms, hydraulic barriers, parking control systems
  - Forced-entry-ballistic-resistant doors, windows, walls, and roofs

  - Barrier protection for man-passable openings (greater than 96 square inches) such as air vents, utility openings, and culverts
  - Mechanical locking systems
  - Elimination of hiding places
  - Multiple layer protection processes
- Perimeter intrusion detection systems
  - Clear zone
  - Video and CCTV surveillance technology
  - Alarms
  - Detection devices (motion, acoustic, infrared)
- Personnel identification systems
  - Access control, fingerprints, biometrics, ID cards
  - Credential management
  - Tailgating policies
  - Primary and secondary credential systems
- Protection of information and data
  - Acoustic shielding
  - Shielding of electronic security devices from hostile electronic environments
  - Computer screen shields
  - Secure access to equipment, networks, and hardware, for example, satellites and telephone systems

Insider Threats

One of the most serious threats may come from persons who have authorized access to a facility. These may include disgruntled employees or persons who have gained access through normal means (e.g., contractors, support personnel, etc.). To mitigate this threat, some items to consider include the following:

- Implement personnel reliability programs and background checks
- Limit and control access to sensitive areas of the facility
- Compartmentalization within the building/campus
- Two-man rule for access to restricted areas
- Video and CCTV surveillance technology

Ballistic Threats

These threats may range from random drive-by shootings to high-powered rifle attacks directed at specific targets within the facility (assassinations). It is important to quantify the potential risk and to establish the appropriate level of protection.

- Obscuration or concealment screening using trees and hedges, berms, solid fencing, walls, and less critical buildings
- Ballistic resistant rated materials and products

- Locating critical assets away from direct lines of sight through windows and doors
- Minimize number and size of windows
- Physical energy absorption screens such as solid fences, walls, and earthen parapets
- Provide opaque windows or window treatments such as reflective coatings, shades, or drapes to decrease sight lines
- Avoid sight lines to assets through vents, skylights, or other building openings
- Use foyers or other door shielding techniques to block observation through a doorway from an outside location
- Avoid main entrances to buildings or critical assets that face the perimeter or an uncontrolled vantage point

WMDs: Chemical, Biological, and Radiological

Commonly referred to as WMD, these threats generally have a low probability of occurrence, but the consequences of an attack may be severe. These threats may be delivered by hand, mail, or as a result of accidental release of toxic industrial agents. While fully protecting a facility against such threats may not be feasible with few exceptions, there are several common sense and low-cost measures that can improve resistance and reduce the risks. Some items to consider include the following:

- Protect ventilation pathways into the building
  - Control access to air inlets and water systems
  - Locate air intake well above ground level
  - Provide detection and filtration systems for HVAC systems, air intakes and water systems
  - Provide for emergency HVAC shutoff and control
  - Segregate portions of building spaces (i.e., provide separate HVAC for the lobby, loading docks, and the core of the building)
  - Consider positive pressurization to keep contaminates outside of the facility
- Provide an emergency notification system to facilitate orderly response and evacuation
- Avoid building locations in depressions where air could stagnate
- Provide access control to mechanical rooms
- Provide CBR monitoring apparatus

Cyber and Information Security Threats

Businesses rely heavily on the transmission of, storage of, and access to a wide range of electronic data and communication systems. Protecting these systems from attack is critical. Some items to consider include the following:

- Understand and identify the information assets you are trying to protect. These may include personal information, business information such as proprietary designs or processes, national security information, or simply the ability of your organization to communicate via email and other LAN/WAN and wireless functions.
- Protect the physical infrastructure that supports information systems. If the computer system is electronically secure but vulnerable to physical destruction, it may need more protection.
- Provide software and hardware devices to detect, monitor, and prevent unauthorized access to or the destruction of sensitive information.

The closing comments made by a person or persons making a terrorist threat, verbally or in writing, may or may not be real, and the infrastructure and its representatives must take that threat seriously and follow the proper protocols as outlined in the organization's emergency threat plan. A terrorist threat needs to be handled with a reasonable conclusion that it may occur and a conclusion may be made that the person's actions convey a threat to commit a violent act. These issues need to be handled with diligence and established guidelines in response to the emergency plan, along with consultation with the intelligent analysis section through the proper chain of command to ensure whether the threat is real, and the most important part is to never underestimate the threat.

12 Mass Casualty Incident and Mass Fatality Incident

A mass casualty incident (often shortened to MCI and sometimes called a multiple-casualty incident or multiple-casualty situation) is any incident in which emergency medical service (EMS) resources, such as personnel and equipment, are overwhelmed by the number and severity of casualties. For example, an incident where a two-person crew is responding to a motor vehicle collision with three severely injured people could be considered an MCI. The general public more commonly recognizes events such as building collapses, train and bus collisions, earthquakes, and other large-scale emergencies as MCIs. Events such as the Oklahoma City bombing in 1995 and the September 11 attacks in 2001 are well-publicized examples of MCIs.

DECLARATION OF AN MCI

An MCI will usually be declared by the first arriving unit at the scene of the incident. However, it may alternately be declared by a dispatcher, based on the information available from people who call their local emergency telephone number about the incident. A formal declaration of an MCI is usually made by an officer or chief of the agency in charge.

Initially, the senior paramedic at the scene will be in charge of the incident, but as additional resources arrive, a senior officer or chief will take command, usually using an incident command system structure to form a unified command to run all aspects of the incident. In the United States, the Incident Command System is known as the National Incident Management System (NIMS). According to the Federal Emergency Management Agency (FEMA), "NIMS provides the template for the management of incidents."

AGENCIES AND RESPONDERS

An MCI can involve a variety of responders and agencies. The most common are listed below.

Emergency Medical Services

Certified first responders or emergency medical responders may arrive as part of local EMS or may arrive on their own. They will assist with all aspects of patient care, including triage and treatment at the scene and transport from the scene to the hospital.

Paramedic and emergency medical technician (EMT) personnel may arrive in ambulances, in their personal vehicles, or from another agency. They will have control of all aspects of patient care, as assigned by the medical officer or incident commander.

Ground ambulances will be assigned to the transport sector to transport patients and personnel to and from the incident scene, emergency departments of hospitals, and a designated helipad. These ambulances may be municipal services, volunteer services, or from private corporations.

Air ambulances will transport patients from the scene or from designated helipads to receiving hospitals.

Fire and Rescue

Firefighters will perform all initial rescue-related operations, as well as fire suppression and prevention. They may also provide medical care if they are trained and assigned to do so. They may arrive on a fire truck, in their personal vehicles, or from another agency. Many areas near airports will have automatic mutual aid agreements with airport fire departments in the event of a plane crash outside of the airport boundaries.

Public Safety

Police officers will secure and control access to the scene, to ensure safety and smooth operations. Utility services will ensure that utilities in the area are turned off as necessary, in order to prevent further injury or damage at the scene.

Specialized Teams

HazMat specialists in Level II/B protection suits carry a patient out of the incident zone to be decontaminated. Specialized rescue teams may be part of the local fire department; they may be associated with the state, provincial, or federal governments; or they may be privately operated teams. These teams are specialists in specific types of rescue, such as urban search and rescue or confined space rescue.

HazMat teams are responsible for cleaning up and neutralizing any hazardous materials at the scene. Sometimes, these will be specialized CBRNE (chemical, biological, radiological, nuclear, and high-yield explosives) teams. National Guard Units have medics specifically trained in mass casualty triage who may be called in to respond to a disaster-related incident.

Public Services

Railways and transportation agencies will be notified if an incident involves their tracks or right-of-way or if they are required to cease operations in and through affected areas. Transportation agencies will provide buses to transport lightly injured people to the hospital. Buses can also provide shelter at the scene (e.g., "warming buses") if required.

The media plays an important role in keeping the general public informed about the incident and in keeping them away from the incident area. However, a public information officer should be assigned as the only designated responder who communicates with the media, to prevent the spread of misinformation.

Nongovernmental organizations such as St. John Ambulance, the Order of Malta, the Red Cross, the Red Crescent, the Medical Reserve Corps, and the Salvation Army will provide assistance with all aspects of an MCI, including trained medical staff, vehicles, individual registration and tracking, temporary shelter, food service, and many other important services.

Hospitals

Hospitals with emergency departments will have an MCI protocol which they initiate as soon as they are notified of an MCI in their community. They will have preparations in place to receive a massive number of casualties, like calling in more staff, pulling extra and spare equipment out of storage, and clearing nonacute patients out of the hospital. Some hospitals will send doctors to the scene of the incident to assist with triage, treatment, and transport of injured persons to the hospital.

This is not an exhaustive list, and many other agencies and groups of people could be involved in an MCI.

FLOW OF AN MCI

Ideally, once an MCI has been declared, a well-coordinated flow of events will occur, using three separate phases: triage, treatment, and transportation.

Triage

The first-arriving crew will conduct triage. Prehospital emergency triage generally consists of a check for immediate life-threatening concerns, usually lasting no more than one minute per patient. In North America, the START system (Simple Triage and Rapid Treatment) is the most common and is considered the easiest to use. Using START, the medical responder assigns each patient to one of four color-coded triage levels, based on his or her breathing, circulation, and mental status.

The triage levels are as follows:

- *Immediate:* Patients who have major life-threatening injuries but are salvageable given the resources available
- *Delayed:* Patients who have non-life-threatening injuries but are unable to walk or exhibit an altered mental status
- *Walking wounded:* Patients who are able to ambulate out of the incident area to a treatment area
- *Deceased or expectant:* Used for victims who are dead or whose injuries make survival unlikely

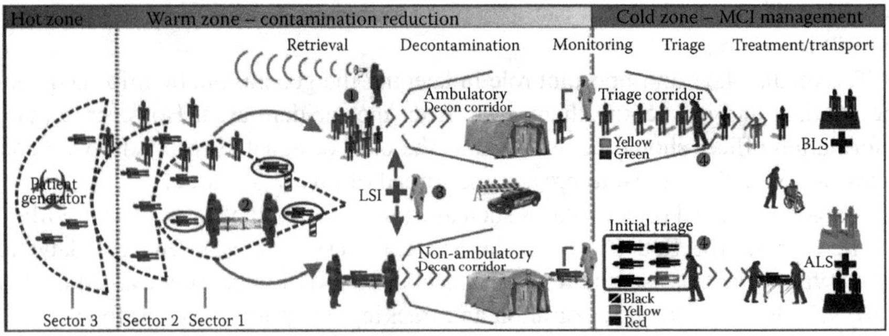

FIGURE 12.1 Simplified HazMat MCI response work flow. ALS, advance life saving; BLS, basic life saving.

Triage personnel do not conduct treatment, with the exception of

- Airway maneuvers;
- Tourniquets for life-threatening hemorrhage; and
- Where allowed by local protocols, needle decompressions for tension pneumothoraces.

Generally, a small group of responders, usually the first two or three crews on scene, can complete triage.

When responding to a chemical, biological, or radiological incident, the first-arriving crew must establish safety zones prior to entering the scene.

Safety zones include the following:

- The hot zone: the contaminated area
- The warm zone: the area where HazMat specialists will decontaminate patients and fellow responders
- The cold zone: the safe zone, where any personnel who are not specially trained in HazMat and do not have chemical or biological protection gear must remain at all times. Depending on the contaminant, the cold zone should be roughly 200–300 yards from the incident. It should also be at least 50 yards uphill and upwind from the warm zone.

These zones should be clearly identified with engineer tapes, lights, or cones. All responders and patients must leave the hot zone in designated pathways into the warm zone, where they will be decontaminated. A designated officer should be posted at the hot zone and warm zone to make sure all contaminated personal are treated and decontaminated before entering the cold zone (Figure 12.1).

Treatment

Once casualties have been triaged, they can be moved to appropriate treatment areas. Unless a patient is Green Tagged and ambulatory, litter bearers will have to transport patients from the incident scene to safer treatment areas located nearby. These treatment areas must always be within walking distance and will be staffed by appropriate numbers of properly certified medical personnel and support people. The litter bearers do not have to be advanced medical personnel; their role is to simply place casualties onto carrying devices and transport them to the appropriate treatment area. Casualties should be transported in order of treatment priority: red-tagged patients first, followed by yellow tagged, then green tagged, and finally black tagged.

Each colored triage category will have its own treatment area. Treatment areas are often defined by colored tarpaulins, flagging tape, signs, or tents. Upon arrival in the treatment area, the casualties are reassessed and they are treated with the goal of stabilizing them until they can be transported to hospitals, transported to the morgue or medical examiner's office, or released.

On-Site Morgue

Some MCIs require an on-site morgue, for several reasons:

- To await transfer of these victims to a permanent morgue
- When the deceased must be removed to access injured victims
- To keep the deceased out of public sight and prevent heightening distress, fear, or panic in an already emotionally-charged scene
- Most often, on-site morgues are set up on the far side of the incident and is in an enclosed area such as a temporary tent or nearby building
- Transport
- Ambulance on scene with emergency lights on

The final stage in the prehospital management of a MCI is the transport of casualties to hospitals for more definitive care. If an insufficient number of ambulances are available, other vehicles may transport patients, such as police cars, fire trucks, air ambulances, transit buses, or personal vehicles. As with treatment, transport priority is decided based on the severity of the patient's injuries. Usually, the most seriously injured are transported first, with the least serious transported only after all the critical patients have been transported.

However, in an effort to remove as many lightly injured civilians as possible, an incident commander may choose to have those least seriously injured transported to local hospitals or interim-care centers in order to provide more room for emergency personnel to work. It is also possible that lightly injured casualties will be transported first when access to those who are more severely injured will be delayed due to heavy or difficult rescue efforts.

DEFINITIVE CARE

The care that is rendered at the scene of an MCI is usually only temporary and is designed to stabilize the casualties until they can receive more definitive care at a hospital or an interim-care center.

Interim-Care Center

An interim-care center is a temporary treatment center that allows for the assessment and treatment of patients until they can either be discharged or transported to a hospital. These are often placed in gymnasiums, schools, arenas, community centers, hotels, and/or other locations that can support a field hospital setup. Permanent buildings are preferred to tents as they provide shelter, power, and running water, but many governments maintain complete field hospital setups that can be deployed anywhere within their jurisdiction within 12–24 hours. While full field hospitals require a significant amount of time to deploy (in relation to the length of most incidents), emergency personnel can set up temporary interim-care centers fairly quickly if needed using the personnel and resources they have on-hand. These centers are usually staffed by a combination of doctors, nurses, paramedics/EMTs, first responders, and social workers (e.g., from the Red Cross), who work to get families reunited after a disaster.

MASS CASUALTY EVENT

Generally, in the healthcare field, the term "mass casualty event" is used when hospital resources are overwhelmed by the number or severity of casualties. During these incidents, hospitals can discharge all fit patients, dedicate more resources to the emergency department, and expand their intensive care unit to accommodate anticipated long-term care needs. While up to 80% of victims will be transported from the scene to hospitals, others who are less injured might walk themselves to these facilities and increase the load at the closest facility to the incident.

MCEs can include bioterrorism attacks, chemical emergencies, radiation emergencies, and natural disasters like weather.

MASS FATALITY INCIDENT

Mass Fatality Definition

A mass fatality incident (MFI), by definition, is any situation where more deaths occur than can be handled by local medical examiner/coroner (ME/C) resources. There is no minimum number of deaths for an incident to be considered an MFI because communities vary in size and resources.

An MFI may be caused by natural hazards (e.g., earthquakes, floods, and hurricanes), human-related hazards (e.g., airline accidents and bridge or tunnel collapses), and proactive human hazards (e.g., terrorist acts).

A mass fatality becomes a catastrophic mass fatality when, as defined in California (Supervisors, 2012), the "loss of life overwhelms the state's mutual aid system and requires extraordinary support from state, federal, and private resources." This definition may vary from state to state. However, in all states, a catastrophic mass fatality is likely to trigger disaster declarations at the state and federal level, and a federal disaster declaration will mobilize an array of resources to support state and local response and recovery efforts.

Regardless of the size of the MFI, the ME/C is the legal authority to conduct victim identification (or augment the lead investigative agencies to complete victim identification), determine the cause and manner of death, and manage death certification. The ME/C is also responsible for other medico-legal activities, such as notification of next of kin.

MFI is an emergency management term used to identify an incident involving more dead bodies and/or body parts than can be located, identified, and processed for final disposition by available response resources.

Although it is a somewhat relative term in that there is no widely accepted number of fatalities that define an MFI, it is generally recognized that if the number of fatalities exceeds the local city or county's resource capabilities, causing them to request assistance or mutual aid from neighboring jurisdictions, the term applies.

MFIs may or may not be a result of an MCI, which is considered a different type of incident and usually focuses more on managing the surviving victims of an incident. MFI and MCI may, and often do, occur simultaneously. MFI differs from MCI in that most, if not all, of the victims of the incident are deceased. A catastrophic plane crash with no survivors is one example of an MFI. Part of the distinction between MFI and MCI is because different kinds of resources are needed to manage each. Living victims are attended to by medical personnel such as EMS, and deceased victims are attended to by medical examiners or coroners.

MFIs may be either human caused, such as hazardous materials releases, transportation accidents, or terrorist attacks, or they may be the result of natural disasters such as earthquakes, floods, or severe weather.

Some significant MFIs are

- South Asian tsunami disaster
- Atomic bombings of Hiroshima and Nagasaki
- Hurricane Katrina
- The September 11 attacks
- Oklahoma City bombing
- Bhopal disaster
- Spanish flu

One of the naturally occurring incidents with great potential to cause a MFI is pandemic influenza (flu). The Spanish Flu pandemic of 1918 killed millions and overwhelmed response resources on a global level. A modern pandemic could have similarly overwhelming impacts. Catastrophic incidents that result in mass fatalities usually also result in mass injuries and/or illnesses. While it is more important to dedicate resources to care for the living, many people have public health concerns about the dead. This is one important reason why jurisdictions usually include mass fatality planning as part of their overall emergency preparedness efforts.

After some MFI authorities have conducted hasty mass burials, research has shown that this is a generally unsatisfactory response.

Mass burials are usually not required for public health reasons, as they increase distress among survivors and interfere with long-term community recovery. In all cultures, there are customs and rituals for dealing with the dead. Universally, survivors want to know what happened to their loved ones and that their loved ones' remains were treated with respect. These are important reasons to plan for mass fatality management.

Response Functions

The primary response functions in a MFI are the following:

- Human remains recovery—the search and rescue efforts to locate bodies and body parts, marking and documenting the location of found remains and eventually transporting the remains to either decontamination or the site morgue for examination as appropriate.
- Decontamination (depending on event)—the "cleaning" of either chemically or biologically contaminated remains to make them safe for further handling and examination.
- Examination
- Identification and death certification
- Processing for final disposition

Mass Fatality Management Resources

The following may provide helpful information for mass fatality planning: www.dmort7.org. The Disaster Mortuary Operational Response Team (DMORT) is part of the National Disaster Medical System (NDMS) and provides support to the National Transportation Safety Board and other mass fatality requirements (NACCHO, 2016). The DMORT is "A free, online collection of local public health tools produced by members of the public health community" (NACCHO, 2017).

- Section A—Acknowledgements
- Section B—Introduction
- Section C—Planning Context
- Section D—Concept Of Operation
- Section E—Incident Notification/Plan Activation
- Section F—Command and Control
- Section G—Human Remains Recovery
- Section H—Morgue Services
- Section I—Family Assistance
- Section J—Public Communications
- Section K—Death Certificates and Disposition Permits
- Section L—Death Care Industry
- Section M—Mass Fatality Plan Maintenance
- Section N—Security
- Section O—Mass Fatality Information Systems
- Section P—Staff/Volunteer Processing Center
- Section Q—Family Concerns and Religious/Cultural Considerations
- Section R—Infection and Other Health and Safety Threats
- Section S—Pandemic Influenza

DISASTER MORTUARY OPERATIONAL RESPONSE TEAM

A Disaster Mortuary Operational Response Team, or DMORT, is a team of experts in the fields of victim identification and mortuary services. DMORTs are activated in response to large-scale disasters in the United States to assist in the identification of deceased individuals and storage of the bodies pending the bodies being claimed.

ORGANIZATION

For organizational purposes, the country is divided into 10 regions, each with a regional coordinator. For the duration of their service, DMORT members work under the local authorities of the disaster site and their professional licenses are recognized by all states.

DMORT Teams:

- REGION I (ME, NH, VT, MA, CT, RI)
- REGION II (NY, NJ, PR, VI)
- REGION III (PA, MD, DC, DE, VA, WV)
- REGION IV (AL, KY, TN, NC, SC, GA, MS, FL)
- REGION V (MN, WI, IL, IN, MI, OH)
- REGION VI (NM, TX, OK, AR, LA)
- REGION VII (NE, IA, KS, MO)
- REGION VIII (MT, ND, SD, WY, UT, CO)
- REGION IX (AZ, NV, CA, HI)
- REGION X (WA, AK, OR, ID)

MORTs are organized under the Department of Health and Human Services (HHS) NDMS. The DMORTs are composed of civilian funeral directors, medical examiners, coroners, pathologists, forensic anthropologists, fingerprint specialists, forensic odontologists, dental assistants, and radiographers. They are supported by medical records technicians and transcribers, mental health specialists, computer professionals, administrative support staff, and security and investigative personnel. When a DMORT is activated, the personnel on the team are treated and paid as a temporary federal employee.

The HHS maintains three Disaster Portable Morgue Units (DPMUs), which are staged at HHS Logistics Centers, one each in Frederick, Maryland, Fort Worth, Texas, and San Jose, California. Each DPMU is a cache of equipment and supplies for a complete morgue with designated workstations for each process the DMORT team is required to complete.

IDENTIFICATION OF REMAINS

Identification of remains is a two-part process that utilizes a sophisticated computer program for matching physical characteristics. The families of the deceased provide as much information about them as possible: dental records, X-rays, photographs or descriptions of tattoos, clothing and jewelry, blood type information, and objects that may contain the deceased's DNA, such as hair or a toothbrush.

The information gathered, called antemortem, or "before death," information is entered into a computer program called VIP (Victim Identification Profile), which is capable of assimilating 800 different item categories, including graphics, photographs, and X-rays. As forensic scientists (pathologists, anthropologists, and odontologists) examine the recovered remains, they enter their findings—called postmortem data—into VIP. Depending on the availability of data, the Windows-VIP system enables scientists to match the remains to their identity.

Incidents

For the World Trade Center disaster, U.S. Secretary of Health and Human Services Tommy G. Thompson activated the NDMS. It was the first time this federally coordinated response system had been activated on a full nationwide basis.

In 2006, DMORT operated the Find Family National Call Center in Baton Rouge, Louisiana. This is the center of all operations concerning the location and reuniting of families scattered by Hurricane Katrina and Hurricane Rita. Out of nearly 13,000 people reported missing after the impacts of hurricanes Katrina, Rita, Stan, and Wilma, nearly 7,000 were found alive and reunited with their families.

MASS FATALITY MANAGEMENT

The ME/C is responsible for the recovery, identification, and disposition of MFI victims.

One of the sad realities of disasters is that they can result in the loss of human life. Many emergency responders have learned this from their own experience, and they will remember the incident for a long time. Mass fatality disasters are often long remembered by the responders, the community, the state, and the nation. Disaster management is important, and every plan should have a section on fatality management.

When a passenger aircraft crashes carrying 100 to 300 passengers or a building structure with hundreds of people inside collapses, the incident would probably be deemed an MFI, even in the largest of cities. Likewise, an incident like a mad gunman killing five people can be called a multiple fatality incident, but it may not be a MFI if it occurs in a major city. In any case, the recovery, identification, and disposition of the victims are the responsibilities of the ME/C of the jurisdiction where the incident occurs.

There are three major operational areas in a mass fatalities incident response:

1. Search and recovery (S&R)
2. Morgue operation
3. Family assistance

Trained personnel should oversee the tasks associated with each of the three areas.

Search and Recovery

Simply stated, S&R normally involves locating and removing at least bodies, body parts, and personal effects. A good S&R team will document everything found at the disaster site, as it may help in the investigation and in the morgue operations. A good policy is to treat every site as a crime scene, until the ME/C says differently. As a rule of thumb, search teams systematically search and mark where bodies, body parts, and personal effects are located with pin flags, stakes, etc.

A team member will assign a number to that particular finding. They log each finding on a grid chart, photograph it, and move on until the search is completed.

Recovery starts after the search of an area is complete. Bodies and body parts must be treated with dignity and respect at all times. Each finding should be tagged with the number assigned by the search team. Bodies and body parts should be placed into a body bag or acceptable substitute. A tag with the same number as the finding inside the bag will be placed on the outside of bag. The body bag should be removed from the scene and taken to a location designated by the ME/C. Personal effects found on the body should not be removed from the victims at the scene. If weathering may be a problem, the personal effects can be wrapped in plastic and affixed to the body or body part. Victim identification is a function within the morgue operations, not the S&R team.

Morgue Operations

Depending on the size and nature of the incident, the ME/C will determine where to establish an incident morgue site. The site may be in the existing morgue for that geographical area or it may be a temporary incident morgue site in another location, such as a warehouse, airplane hangar, or fairground building. School gymnasiums should not be used, particularly when school is in session.

The ME/C should lay out the morgue operation site considering the physical condition of the victims, the number of victims, and the number of personnel needed to perform such morgue functions as administration, logistics, refrigeration, and operations. The operational areas can include areas for receiving, photography, X-ray, personal effects, anthropology, dental, fingerprinting, pathology, storage, and shipping. In some cases, an area for embalming may be desirable.

The main purposes for the morgue are to determine the cause of death and identify victims. The use of highly skilled professionals for each of the morgue operational areas is important. Postmortem records will be completed for every body and body part as they are processed through each of the operational stations. Postmortem records include personal effects, photography, radiographs, anthropology, finger prints, and dental and pathology reports.

The postmortem records will be compared to the antemortem (predeath) records obtained from the victim's family and other sources such as fingerprint repositories and hospitals. Personal effects, such as driver licenses found on the victim or statements of recognition, should not be used as positive identification, but rather tentative identification.

Positive identification is a responsibility of the ME/C. After identification is established, the medical examiner can release the body and/or body parts based on the desires of the "next of kin."

FAMILY ASSISTANCE CENTER

The family assistance center (FAC) is one of the most sensitive operations in a mass fatalities event. Its purposes are as follows:

- To provide relatives of victims with information and access to services they may need in the days following the incident
- To protect families from the media and curiosity seekers
- To allow investigators and the ME/C access to families so they can obtain information more easily

An FAC should be established quickly, in an area such as a hotel, conference center, school, or church. The area selected should be secured, in order to give privacy to the families. Regular briefing by the ME/C or staff twice daily will help keep the families informed. Meeting with the families on an individual basis early on makes it possible to start the process of collecting antemortem records for use in the morgue operations. The FAC has become so important that federal law recommends one to be established whenever a major aviation disaster occurs. Staffing for the FAC is important. Grief counselors should be available. Personnel from the American Red Cross possessing trained counseling skills and funeral service personnel are good at working with grieving families. Translators may be necessary when working with families from foreign countries.

There are many volunteer organizations and community businesses able to assist the ME/C during an MFI response. If you have or are contemplating forming a response team, be sure they receive quality training on the basic MFI response principles and procedures. Contact your local emergency management office for training information or contact your state emergency management training officer to obtain information on scheduling a MFI response course in your area. The course is available from FEMA's Emergency Management Institute and is entitled Mass Fatalities Incident Response Course.

Your response program should include policies and procedures that meet the approval of the ME/C.

They should also be consistent with the overall disaster response plan managed by the emergency management office.

The subject of fatality management must be taken seriously in this day of natural and manmade disasters.

Fatality Management

Fatality management is the ability to coordinate with other organizations (e.g., law enforcement, healthcare, emergency management, and ME/C) to ensure the proper recovery, handling, identification, transportation, tracking, storage, and disposal of human remains and personal effects; certify cause of death; and facilitate access to mental/behavioral health services to the family members, responders, and survivors of an incident.

This capability consists of the ability to perform the following functions:

1. Determine role for public health in fatality management
2. Activate public health fatality management operations
3. Assist in the collection and dissemination of ante mortem data
4. Participate in survivor mental/behavioral health services
5. Participate in fatality processing and storage operations

Function 1: Determine Role for Public Health in Fatality Management

Coordinate with the lead jurisdictional authority (e.g., coroner, medical examiner, sheriff, or other agent) to identify the roles and responsibilities of jurisdictional public health entities in fatality management activities.

This function consists of the ability to perform the following tasks:

Task 1: Prior to an incident, characterize potential fatalities based on jurisdictional risk assessment and the impact of these potential fatalities on jurisdictional resource needs.

Task 2: Prior to an incident, coordinate with subject matter experts (e.g., those with expertise in epidemiology, laboratory, surveillance; community cultural/religious beliefs or burial practices; chemical, biological, radiological, and emergency operations leads; and partners from hospital, mortuary, EMS) to determine public health's role in an incident that may result in fatalities.

Task 3: Prior to an incident, coordinate with jurisdictional, private, and federal Emergency Support Function 6 and Emergency Support Function 8 resources as necessary to determine their roles and requirements for the response.

Function 2: Activate Public Health Fatality Management Operations

This is in accordance with public health jurisdictional standards and practices and as requested by the lead jurisdictional authority.

This function consists of the ability to perform the following tasks:

Task 1: Assess data from the incident to inform and guide the public health resources needed for the response.

Task 2: Identify and coordinate with jurisdictional, regional, private, and federal Emergency Support Function 8 resources with expertise in the potential cause(s) of fatalities to make recommendations regarding all phases of human remains disposition: recovery, processing (e.g., decontamination, infection control, and other mitigation measures), storing, and disposing.

Task 3: Coordinate with partners to initiate predetermined (e.g., local, regional, state, federal, and private sector) processes for all phases of human remains disposition.

Task 4: Coordinate incident details among members of the public health and medical health systems by sharing information between programs and linking information databases, based on the scope of the incident. (For additional or supporting detail, see Capability 6: Information Sharing.)

Function 3: Assist in the Collection and Dissemination of Antemortem Data

Assist, if requested, the lead jurisdictional authority and jurisdictional and regional partners to gather and disseminate ante mortem data 71 through a Federal Acquisition Circular (FAC) Model 72 or other mechanism.

This function consists of the ability to perform the following tasks:

Task 1: Coordinate with partners for the establishment of a mechanism (e.g., FAC) to collect antemortem data.

Task 2: Coordinate with partners to identify and assemble the resources required to collect and communicate antemortem data.

Task 3: Coordinate with partners and assist, if needed, in the collection and dissemination of antemortem data to families of the deceased and law enforcement officials.

Task 4: Coordinate with partners to support electronic recording and reporting of antemortem data through electronic systems and/or other information sharing platforms.

Function 4: Participate in Survivor Mental/Behavioral Health Services

Coordinate with the lead jurisdictional authority and jurisdictional and regional partners to support the provision of non-intrusive, culturally sensitive mental/behavioral health support services to family members of the deceased, incident survivors, and responders, if requested.

This function consists of the ability to perform the following tasks:

Task 1: Coordinate with partners to assemble the required staff and resources to provide nonintrusive mental/behavioral health services to responders.

Task 2: Coordinate with partners to facilitate availability of culturally appropriate assistance (e.g., addressing language barriers and religious or cultural practices).

Task 3: Coordinate with Emergency Support Function 8 partners to support the provision of mental/behavioral health services to family members of the deceased and incident survivors as needed.

Function 5: Participate in Fatality Processing and Storage Operations

Assist the lead jurisdictional authority and partners in ensuring that human remains and associated personal effects are safely recovered, processed, transported, tracked, stored, and disposed of or released to authorized person(s), if requested.

This function consists of the ability to perform the following tasks:

Task 1: Make recommendations to the incident management/jurisdictional lead agency on procedures for the safe recovery, receipt, identification, decontamination, transportation, storage, and disposal of human remains. Recommendations can also include an assessment of the need for temporary burial, procurement of public property for temporary burial, and security/privacy requirements of the processing facility.

Task 2: Assist, if needed or requested, in multispecialty forensic analysis to identify human remains and determine the cause and manner of death.

Task 3: Coordinate with partners to support electronic death reporting.

Task 4: Coordinate with partners to facilitate the collection and reporting of mortality information (e.g., vital records).

SECURITY AT A MASS CASUALTY AND MASS FATALITY INCIDENT

Security is critical to effective mass fatality management. A lack of security can derail the best plans.

In the event of a mass fatality, the incident site, incident morgue, and the FAC will require security and traffic control. The law enforcement agency for the jurisdiction where the incident takes place will be responsible for these operations. When the emergency operations center is activated, the law enforcement branch in the operations section will provide oversight and coordination of law enforcement mutual aid that is called in to assist with site security and traffic control.

A written security plan and traffic control plan that outline the procedures and requirements of the operation are recommended for the incident site, incident morgue, and the FAC during the early stages of mass fatality operations.

Many law enforcement agencies will have established protocols and procedures for developing security and traffic control plans. However, when the ME/C office requests security support for its mission, it needs to be appropriately prepared.

That means being prepared to request the right number and type of security resources that will be needed and being able to promptly provide law enforcement with the information it will need to achieve success in its support mission.

Overview

Mass fatality security requirements, security objectives, and information to request when selecting sites for mass fatality operations are described below. This information is followed by a description of the associated tools included with this section. The associated tools are a questionnaire that can be used to perform a physical security assessment, a security plan template, and a traffic control plan template. These resources have been developed to assist the ME/C office and the local law enforcement agency in effective mass fatality management.

These associated tools are

- Identification of stakeholders involved in mass fatality management and recommendations for a planning process that is co-led by public health and the medical examiner/coroners office;

- Specific guidelines for all medical examiner/coroner responsibilities at the incident site, morgue, and family assistance center;
- Guidance on infection and other health and safety threats; and
- Requirements and recommendations for managing mass fatalities during a worst-case scenario pandemic influenza.

Security Requirements

In a mass fatality, the nature of the incident will dictate required security. The following are general security requirements to anticipate:

- Incident site
- Morgue (jurisdiction's morgue, temporary incident morgue, and long-term examination center/sifting site, if one is required)
- FAC(s)/reception center(s)
- Traffic control at all sites

In addition, agencies tasked to provide security for mass fatality operations may also be tasked with providing heightened security throughout the jurisdiction.

Security Objectives

Key considerations for securing all mass fatality operations sites are the following:

- Controlling access into, within, and out of the facility
- Perimeter protection
- Parking lot protection
- Traffic control
- Establishing and protecting landing zones as needed (e.g., for the delivery of Disaster Mortuary Operational Response Team's portable morgue unit)
- Crowd control

The following are the general security objectives for each site.
Incident site objectives are as follows:

- To control site access 24/7
- All authorized personnel, volunteers, and approved visitors must have photo ID security badges that reference function and access
  - To recover human remains
  - To preserve evidence
- To facilitate identification of victims
- If a crime or terrorism is suspected
  - To protect response personnel and volunteers
  - To protect the public from potential physical dangers (e.g., building collapse) and when chemical, biological, and/or radiological agents are involved

- To escort vehicles transporting human remains from the incident site to the morgue (as needed)
- To secure parking areas

The morgue services objectives are as follows:

- To control site access 24/7
- All authorized personnel, volunteers, and approved visitors must have photo ID security badges that reference function and access
- To preserve evidence
- To facilitate identification of victims
- If a crime or terrorism is suspected
- To protect morgue personnel and volunteers
- To protect the deceased
- To secure parking areas

FAC objectives are as follows:

- To control site access 24/7
- All authorized personnel, volunteers, and approved visitors must have photo ID security badges that reference function and access
- All family members and loved ones of potential victims must have photo ID security badges that reference their roll and access
- No media, curiosity seekers, general public, etc., are to be admitted under any circumstances without authorization
- To ensure that families feel protected when visiting the FAC
- May want to consider police in plain clothes who patrol inside the FAC to ensure that no unauthorized persons have gained entry
- To protect response personnel and volunteers
- To secure parking areas

Security is a high priority at the FAC.

Information to Request for Security Planning When Selecting Sites for Incident Morgue and FAC

Other than the incident site itself, the Emergency Operations Center Logistics Section will be responsible for site selection—the incident morgue and the FAC. The following are security-related recommendations when selecting sites:

- Involve law enforcement in site selection so that major security risks can be identified immediately.
- Collect the following information, as available, for each site:
  - Exterior and interior photos
  - Map-wide view

- Map-tight view
- Aerial photo—wide view
- Aerial photo—tight view
- Facility's floor plan diagram
- Parking plan
- Mass transit map

Give this information with the ME/C's proposed floor plans/layouts for the temporary incident morgue and for the FAC to the law enforcement branch to assist them with development of security and traffic control plans.

Overview of Physical Security Assessment

A Physical Security Assessment is included as a tool with this section. It contains the following sections:

Physical security assessment: exterior of the site

- Perimeter
- Lighting
- Parking areas
- Landscaping

Physical security assessment: interior of building(s)

- Doors, windows, and other openings
- Ceilings and walls
- Emergency power system
- Lighting

Physical security assessment: specific security devices, technologies, and machines

- Alarms
- Fire protection
- Utility control points
- Attic, basements, crawl spaces, and air-conditioning and heating ducts
- Communications
- Physical security assessment: roadway access
- Physical security assessment: neighborhood characteristics (within four blocks of the site)
- Physical security assessment: standard operating procedures. Public areas (waiting areas, restrooms, and hallways). Offices within the facility that handle money
- Security procedures

These questions are followed by a space for any specific security concerns and a summary rating system to present an overview of security issues and requirements.

Overview of Security and Traffic Control Plan Templates

The Security Plan Template includes the following sections:

- Security plan staffing and postings
- Security postings, interior
- Security postings, exterior
- Site specific security operations plan and comments
- Diagram and photos of facility utility shut-off controls

The Traffic Control Plan Template includes the following sections:

- Traffic control plan staffing and postings
- Traffic control postings
- Site-specific traffic control operations plan and comments

Once the security and traffic control plans have been completed, standard operating procedures will be needed. It is expected that law enforcement agencies will modify existing applicable operating procedures as required to implement the security and traffic control plans.

The Physical Security Assessment, Security Plan Template, and Traffic Control Plan Template are included as tools to assist jurisdictions in developing mass fatality security and traffic control plans.

These tools are based on the work of many County Department of Public Health Emergency Preparedness & Response Program's multidisciplined Strategic National Stockpile Force Protection Committee's work on security for Points of Dispensing Preplans.

The purpose of mass casualty and mass fatality incidents is to respond to and manage differently than our normal response system (Figure 12.2).

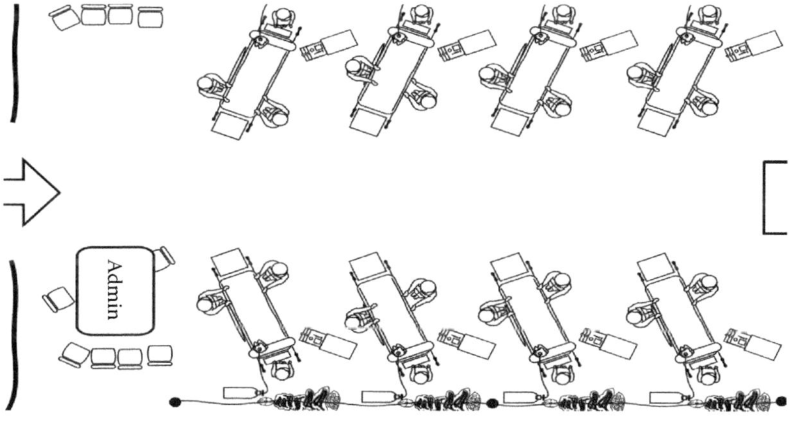

FIGURE 12.2 Example mass fatality layout.

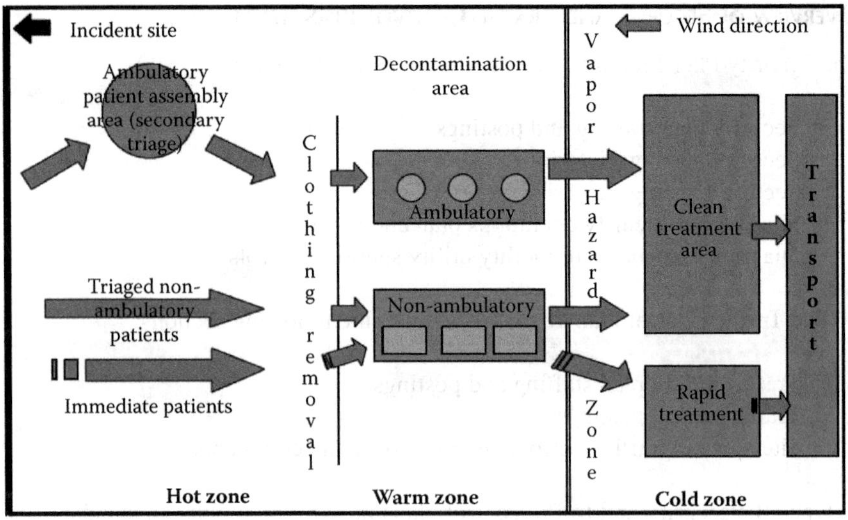

FIGURE 12.3 Example of a mass casualty and mass fatality layout.

Most mass casualty/fatality incidents occur without notice, yet require a major response effort. Because of the number of casualties, first arriving emergency workers are often overwhelmed.

A majority of our existing plans are designed to effectively respond to major incidents. However, for mass casualty/fatality incidents, they must be expanded in scope very quickly (Figure 12.3). The incident management provides the means to accomplish this and maintain control.

13 Emergency Management and Weapons of Mass Destruction

A weapon of mass destruction (WMD) is a nuclear, radiological, chemical, biological, or other weapon that can kill and bring significant harm to a large number of humans or cause great damage to human-made structures (e.g., buildings), natural structures (e.g., mountains), or the biosphere. The scope and application of the term has evolved and been disputed, often signifying more politically than technically. Originally coined in reference to aerial bombing with chemical explosives, since World War II it has come to refer to large-scale weaponry of other technologies, such as chemical, biological, radiological, or nuclear.

DEFINITIONS OF THE TERM

STRATEGIC

The most widely used definition of "weapons of mass destruction" is that of nuclear, biological, or chemical weapons (NBC), although there is no treaty or customary international law that contains an authoritative definition. Instead, international law has been used with respect to the specific categories of weapons within WMD, and not to WMD as a whole.

While nuclear, chemical, and biological weapons are regarded as the three major types of WMDs, some analysts have argued that radiological materials as well as missile technology and delivery systems such as aircraft and ballistic missiles could be labeled as WMDs as well.

The abbreviations NBC (for nuclear, biological and chemical) and CBR (chemical, biological, and radiological) are used with regards to battlefield protection systems for armored vehicles, because all three involve insidious toxins that can be carried through the air and can be protected against with vehicle air filtration systems.

However, there is an argument that nuclear and biological weapons do not belong in the same category as chemical and "dirty bomb" radiological weapons, which have limited destructive potential (and close to none, as far as property is concerned), whereas nuclear and biological weapons have the unique ability to kill large numbers of people with very small amounts of material, and thus could be said to belong in a class by themselves.

The NBC definition has also been used in official U.S. documents, by the U.S. President, the U.S. Central Intelligence Agency, the U.S. Department of Defense, and the U.S. Government Accountability Office.

Other documents expand the definition of WMD to also include radiological or conventional weapons. The U.S. military refers to WMD as (Department of Defense, 2012):

> Chemical, biological, radiological, or nuclear weapons capable of a high order of destruction or causing mass casualties and exclude the means of transporting or propelling the weapon where such means is a separable and divisible part from the weapon. Also called WMD.

This may also refer to nuclear ICBMs (intercontinental ballistic missiles).

The significance of the words *separable* and *divisible part of the weapon* is that missiles such as the Pershing II and the SCUD are considered weapons of mass destruction, while aircraft capable of carrying bomb loads are not.

In 2004, the United Kingdom's (2006) Butler Review recognized the "considerable and long-standing academic debate about the proper interpretation of the phrase 'weapons of mass destruction.'" The committee set out to avoid the general term but when using it employed the definition of United Nations Security Council Resolution 687, which defined the systems that Iraq was required to abandon:

- Nuclear weapons or nuclear-weapons-usable material or any sub-systems or components or any research, development, support or manufacturing facilities relating to [nuclear weapons].
- Chemical and biological weapons and all stocks of agents and all related subsystems and components and all research, development, support and manufacturing facilities.
- Ballistic missiles with a range greater than 150 kilometers and related major parts, and repair and production facilities.

Chemical weapons expert Gert G. Harigel considers only nuclear weapons true WMDs because "only nuclear weapons are completely indiscriminate by their explosive power, heat radiation and radioactivity, and only they should therefore be called a weapon of mass destruction." He prefers to call chemical and biological weapons "weapons of terror" when aimed against civilians and "weapons of intimidation" for soldiers (Harigel, 2017).

For a period of several months in the winter of 2002–2003, U.S. Deputy Secretary of Defense Paul Wolfowitz frequently used the term "weapons of mass terror," apparently also recognizing the distinction between the psychological and the physical effects of many things currently falling into the WMD category.

Gustavo Bell Lemus, the vice president of Colombia, at the July 9, 2001, United Nations (UN) Conference on the Illicit Trade in Small Arms and Light Weapons in All Its Aspects, quoted the Millennium Report of the UN Secretary-General to the General Assembly, in which Kofi Annan said that small arms could be described as WMD because the fatalities they cause "dwarf that of all other weapons systems—and in most years greatly exceed the toll of the atomic bombs that devastated Hiroshima and Nagasaki" (United Nations, 2001).

An additional condition often implicitly applied to WMD is that the use of the weapons must be strategic. In other words, they would be designed to "have consequences far outweighing the size and effectiveness of the weapons themselves."

The strategic nature of WMD also defines their function in the military doctrine of total war as targeting the means a country would use to support and supply its war effort, specifically its population, industry, and natural resources.

Within U.S. civil defense organizations, the category is now Chemical, Biological, Radiological, Nuclear, and Explosive, which defines WMD as

- Any explosive, incendiary, poison gas, bomb, grenade, or rocket having a propellant charge of more than four ounces [113 g], missile having an explosive or incendiary charge of more than one-quarter ounce [7 g], or mine or device similar to the above.
- Poison gas.
- Any weapon involving a disease organism.
- Any weapon that is designed to release radiation at a level dangerous to human life.

Military

For the general purposes of national defense, the U.S. Code defines a WMD as:

> Any weapon or device that is intended, or has the capability, to cause death or serious bodily injury to a significant number of people through the release, dissemination, or impact of:
>
> - Toxic or poisonous chemicals or their precursors
> - A disease organism
> - Radiation or radioactivity

For the purposes of the prevention of weapons proliferation, the U.S. Code defines WMDs as "chemical, biological, and nuclear weapons, and chemical, biological, and nuclear materials used in the manufacture of such weapons."

Criminal (Civilian)

For the purposes of U.S. criminal law concerning terrorism, WMDs are defined as follows (FBI, www.fbi.gov):

- Any "destructive device" defined as any explosive, incendiary, or poison gas—bomb, grenade, rocket having a propellant charge of more than four ounces, missile having an explosive or incendiary charge of more than one-quarter ounce, mine, or device similar to any of the devices described in the preceding clauses
- Any weapon that is designed or intended to cause death or serious bodily injury through the release, dissemination, or impact of toxic or poisonous chemicals, or their precursors
- Any weapon involving a biological agent, toxin, or vector
- Any weapon that is designed to release radiation or radioactivity at a level dangerous to human life

The Federal Bureau of Investigation's definition is similar to that presented above from the terrorism statute:

- Any "destructive device" as defined in Title 18 USC Section 921: any explosive, incendiary, or poison gas—bomb, grenade, rocket having a propellant charge of more than four ounces, missile having an explosive or incendiary charge of more than one-quarter ounce, mine, or device similar to any of the devices described in the preceding clauses
- Any weapon designed or intended to cause death or serious bodily injury through the release, dissemination, or impact of toxic or poisonous chemicals or their precursors
- Any weapon involving a disease organism
- Any weapon designed to release radiation or radioactivity at a level dangerous to human life
- Any device or weapon designed or intended to cause death or serious bodily injury by causing a malfunction of or destruction of an aircraft or other vehicle that carries humans or of an aircraft or other vehicle whose malfunction or destruction may cause said aircraft or other vehicle to cause death or serious bodily injury to humans who may be within range of the vector in its course of travel or the travel of its debris

Indictments and convictions for possession and use of WMD such as truck bombs, pipe bombs, shoe bombs, and cactus needles coated with a biological toxin have been obtained under 18 USC 2332a.

As defined by 18 USC §2332 (a) ("Use of Weapons of Mass Destruction," https://www.law.cornell.edu/uscode/text/18/2332a), a WMD is

- Any destructive device as defined in section 921 of the title.
- Any weapon that is designed or intended to cause death or serious bodily injury through the release, dissemination, or impact of toxic or poisonous chemicals, or their precursors.
- Any weapon involving a biological agent, toxin, or vector (as those terms are defined in section 178 of this title).
- Any weapon that is designed to release radiation or radioactivity at a level dangerous to human life.

TREATIES NOT TO USE WMDs

The development and use of WMD are governed by several international conventions and treaties, although not all countries have signed and ratified them:

- Partial Test Ban Treaty
- Outer Space Treaty
- Nuclear Non-Proliferation Treaty (NPT)
- Seabed Arms Control Treaty
- Comprehensive Test Ban Treaty (CTBT, has not entered into force as of 2015)

- Biological and Toxin Weapons Convention (BWC)
- Chemical Weapons Convention (CWC)

A variety of treaties and agreements have been enacted to regulate the use, development, and possession of various types of WMDs. Treaties may regulate weapons use under the customs of war (Hague Conventions and Geneva Protocol), ban specific types of weapons (Chemical Weapons Convention and Biological Weapons Convention), limit weapons research (Partial Test Ban Treaty and Comprehensive Nuclear-Test-Ban Treaty), limit allowable weapons stockpiles and delivery systems (START I and SORT), or regulate civilian use of weapon precursors (Chemical Weapons Convention and Biological Weapons Convention). The history of weapons control has also included treaties to limit effective defense against WMDs in order to preserve the deterrent doctrine of mutual assured destruction (Anti-Ballistic Missile Treaty) as well as treaties to limit the spread of nuclear technologies geographically (African Nuclear Weapons Free Zone Treaty and Nuclear Non-Proliferation Treaty).

USE, POSSESSION, AND ACCESS

Nuclear Weapons

The only country to have used a nuclear weapon in war is the United States, which dropped two atomic bombs on the Japanese cities of Hiroshima and Nagasaki during World War II.

There are eight countries that have declared they possess nuclear weapons and are known to have tested a nuclear weapon, only five of which are members of the NPT. The eight are China, France, India, North Korea, Pakistan, Russia, the United Kingdom, and the United States.

Israel is considered by most analysts to have nuclear weapons numbering in the low hundreds as well but maintains an official policy of nuclear ambiguity, neither denying nor confirming its nuclear status.

South Africa developed a small nuclear arsenal in the 1980s but disassembled them in the early 1990s, making it the only country to have fully given up an independently developed nuclear weapons arsenal.

Belarus, Kazakhstan, and Ukraine inherited stockpiles of nuclear arms following the break-up of the Soviet Union but relinquished them to the Russian Federation. Countries with access to nuclear weapons through nuclear sharing agreements include Belgium, Germany, Italy, the Netherlands, and Turkey.

U.S. Politics

Due to the indiscriminate impact of WMD, the fear of a WMD attack has shaped political policies and campaigns, fostered social movements, and has been the central theme of many films. Support for different levels of WMD development and control varies nationally and internationally. Yet understanding of the nature of the threats is not high, in part because of imprecise usage of the term by politicians and the media.

An Atomic-Bomb Blueprint

Fear of WMD, or of threats diminished by the possession of WMD, has long been used to catalyze public support for various WMD policies. They include mobilization of pro- and anti-WMD campaigners alike and generation of popular political support. The term WMD may be used as a powerful buzzword or to generate a culture of fear. It is also used ambiguously, particularly by not distinguishing among the different types of WMD.

More recently, the threat of potential WMD in Iraq was used by President George W. Bush as justification for the 2003 invasion of Iraq. Broad reference to Iraqi WMD in general was seen as an element of President Bush's arguments. The claim that Iraq possessed WMDs led to the invasion of Iraq in 2003 by Coalition forces.

Over 500 munitions were discovered throughout Iraq since 2003 containing chemical agents mustard and Sarin gas, produced in the 1980s and no longer usable as originally intended.

In 2004, Polish troops found nineteen 1980s-era rocket warheads, thwarting an attempt by militants to buy them at $5,000 each. Some of the rockets contained extremely deteriorated nerve agent.

The American Heritage Dictionary defines a WMD as "a weapon that can cause widespread destruction or kill large numbers of people, especially a nuclear, chemical, or biological weapon." In other words, it does not have to be NBC.

For example, the terrorist for the Boston Marathon bombings was charged under U.S. law 18 U.S.C. 2332A for using a WMD, and that was a pressure cooker bomb. In other words, it was a weapon that caused large-scale death and destruction, without being an NBC weapon.

MEDIA COVERAGE

In 2004, the Center for International and Security Studies at Maryland (CISSM) released a report examining the media's coverage of WMD issues during three separate periods: nuclear weapons tests by India and Pakistan in May 1998, the U.S. announcement of evidence of a North Korean nuclear weapons program in October 2002, and revelations about Iran's nuclear program in May 2003 (Moeller, 1998). The CISSM report notes that poor coverage resulted less from political bias among the media than from tired journalistic conventions. The report's major findings were that:

> Most media outlets represented WMD as a monolithic menace, failing to adequately distinguish between weapons programs and actual weapons or to address the real differences among chemical, biological, nuclear, and radiological weapons.

Most journalists accepted the Bush administration's formulation of the "War on Terror" as a campaign against WMD, in contrast to coverage during the Clinton era, when many journalists made careful distinctions between acts of terrorism and the acquisition and use of WMD.

Many stories reported the incumbent administration's perspective on WMD, giving too little critical examination of the way officials framed the events, issues, threats, and policy options, and too few stories proffered alternative perspectives to an official line, a problem exacerbated by the journalistic prioritizing of breaking-news stories and the "inverted pyramid" style of storytelling.

In a separate study published in 2005 (Strumfels, 2015), a group of researchers assessed the effects reports and retractions in the media had on people's memory regarding the search for WMD in Iraq during the 2003 Iraq War.

The study focused on populations in two coalition countries (Australia and the United States) and one opposed to the war (Germany). Results showed that U.S. citizens generally did not correct initial misconceptions regarding WMD, even following disconfirmation; Australian and German citizens were more responsive to retractions. Dependence on the initial source of information led to a substantial minority of Americans exhibiting false memory that WMDs were indeed discovered, while they were not.

This led to three conclusions:

- The repetition of tentative news stories, even if they are subsequently disconfirmed, can assist in the creation of false memories in a substantial proportion of people.
- Once information is published, its subsequent correction does not alter people's beliefs unless they are suspicious about the motives underlying the events the news stories are about.
- When people ignore corrections, they do so irrespective of how certain they are that the corrections occurred.

A poll conducted between June and September 2003 asked people whether they thought evidence of WMD had been discovered in Iraq since the war ended.

They were also asked which media sources they relied upon. Those who obtained their news primarily from Fox News were three times as likely to believe that evidence of WMD had been discovered in Iraq as those who relied on PBS and NPR for their news, and one third more likely than those who primarily watched CBS.

NUCLEAR TERRORISM

Nuclear terrorism refers to an act of terrorism in which a person or persons belonging to a terrorist organization detonates a nuclear device. Some definitions of nuclear terrorism include the sabotage of a nuclear facility and/or the detonation of a radiological device, colloquially termed a dirty bomb, but consensus is lacking. In legal terms, nuclear terrorism is an offense committed if a person unlawfully and intentionally "uses in any way radioactive material…with the intent to cause death or serious bodily injury; or with the intent to cause substantial damage to property or to the environment; or with the intent to compel a natural or legal person, an international organization or a State to do or refrain from doing an act," according to the 2005 UN International Convention (https://treaties.un.org/doc/db/Terrorism/english-18-15.pdf) for the Suppression of Acts of Nuclear Terrorism.

The possibility of terrorist organizations using nuclear weapons (including those of a relatively smaller size), such as those contained within suitcases (suitcase nuclear device), is something which is known of within U.S. culture and at times previously discussed within the political settings of the United States. It is considered plausible that terrorists could acquire a nuclear weapon.

However, despite thefts and trafficking of small amounts of fissile material, all low-concern and less than Category III Special nuclear material (SNM), there is no credible evidence that any terrorist group has succeeded in obtaining Category I SNM, the necessary multi-kilogram critical mass amounts of weapons grade plutonium required to make a nuclear weapon.

SCOPE

Nuclear terrorism could include the following:

- Acquiring or fabricating a nuclear weapon
- Fabricating a dirty bomb
- Attacking a nuclear reactor, for example, by disrupting critical inputs (e.g., water supply)
- Attacking or taking over a nuclear-armed submarine, plane, or base

Nuclear terrorism, according to a 2011 report published by the Belfer Center for Science and International Affairs at Harvard University, can be executed and distinguished via four pathways (Harvard Kennedy School of Government, 2007):

1. The use of a nuclear weapon that has been stolen or purchased on the black market
2. The use of a crude explosive device built by terrorists or by nuclear scientists who the terrorist organization has furtively recruited
3. The use of an explosive device constructed by terrorists and their accomplices using their own fissile material
4. The acquisition of fissile material from a nation-state

U.S. President Barack Obama calls nuclear terrorism "the single most important national security threat that we face." In his first speech to the U.N. Security Council, President Obama said that "Just one nuclear weapon exploded in a city—be it New York or Moscow, Tokyo or Beijing, London or Paris—could kill hundreds of thousands of people." It would "destabilize our security, our economies, and our very way of life."

Militant Groups

Nuclear weapons materials on the black market are a global concern, and there is concern about the possible detonation of a small, crude nuclear weapon by a militant group in a major city, with significant loss of life and property.

It is feared that a terrorist group could detonate a dirty bomb, a type of radiological weapon. A dirty bomb is made of any radioactive source and a conventional explosive. There would be no nuclear blast and likely no fatalities, but the radioactive material is dispersed and can cause extensive fallout depending on the material used.

A foot-long stick of radioactive cobalt could be taken from a food irradiation plant and combined with ten pounds of explosives to contaminate 1,000 square kilometers and make some areas uninhabitable for decades. There are other radiological weapons called radiological exposure devices where an explosive is not necessary. A radiological weapon may be very appealing to terrorist groups as it is highly successful in instilling fear and panic among a population (particularly because of the threat of radiation poisoning) and would contaminate the immediate area for some period of time, disrupting attempts to repair the damage and subsequently inflicting significant economic losses.

Incidents Involving Nuclear Material

Information reported to the International Atomic Energy Agency (IAEA) shows "a persistent problem with the illicit trafficking in nuclear and other radioactive materials, thefts, losses and other unauthorized activities" (International Atomic Energy Agency, 2017). The IAEA Illicit Nuclear Trafficking Database notes 1,266 incidents reported by 99 countries over the last 12 years, including 18 incidents involving highly enriched uranium (HEU) or plutonium trafficking:

Some examples of the threats that could have been an issue if successful are the following:

- There have been 18 incidents of theft or loss of HEU and plutonium confirmed by the IAEA.
- Security specialist Shaun Gregory argued in an article that terrorists have attacked Pakistani nuclear facilities three times in the recent past, twice in 2007 and once in 2008.
- In November 2007, burglars with unknown intentions infiltrated the Pelindaba nuclear research facility near Pretoria, South Africa. The burglars escaped without acquiring any of the uranium held at the facility.
- In June 2007, the Federal Bureau of Investigation released to the press the name of Adnan Gulshair el Shukrijumah, allegedly the operations leader for developing tactical plans for detonating nuclear bombs in several American cities simultaneously.
- In November 2006, MI5 warned that al-Qaida was planning on using nuclear weapons against cities in the United Kingdom by obtaining the bombs via clandestine means.
- In February 2006, Oleg Khinsagov of Russia was arrested in Georgia, along with three Georgian accomplices, with 79.5 grams of 89% HEU.
- The Alexander Litvinenko poisoning with radioactive polonium "represents an ominous landmark: the beginning of an era of nuclear terrorism," according to Andrew J. Patterson (Sensagent Corporation, 2016).
- In June 2002, U.S. citizen José Padilla was arrested for allegedly planning a radiological attack on the city of Chicago; however, he was never charged with such conduct. He was instead convicted of charges that he conspired to "murder, kidnap and maim" people overseas.

14 Cyberspace and Emergency Management

Cyberspace is "the notional environment in which communication over computer networks occurs" (Reynolds, 2016). The word became popular in the 1990s when the uses of the Internet, networking, and digital communication were all growing dramatically and the term "cyberspace" was able to represent the many new ideas and phenomena that were emerging.

The parent term of cyberspace is "cybernetics," derived from the Ancient Greek (steersman, governor, pilot, or rudder), a word introduced by Norbert Wiener for his pioneering work in electronic communication and control science.

As a social experience, individuals can interact, exchange ideas, share information, provide social support, conduct business, direct actions, create artistic media, play games, engage in political discussion, and so on, using this global network. They are sometimes referred to as cybernauts.

The term "cyberspace" has become a conventional means to describe anything associated with the Internet and the diverse Internet culture.

The U.S. government recognizes the interconnected information technology (IT) and the interdependent network of IT infrastructures operating across this medium as part of the U.S. national critical infrastructure. Among individuals on cyberspace, there is believed to be a code of shared rules and ethics mutually beneficial for all to follow, referred to as cyber ethics.

Many view the right to privacy as most important to a functional code of cyber ethics. Such moral responsibilities go hand in hand when working online with global networks, specifically, when opinions are involved with online social experiences.

Cyberspace is defined more by the social interactions involved rather than its technical implementation. In their view, the computational medium in cyberspace is an augmentation of the communication channel between real people; the core characteristic of cyberspace is that it offers an environment that consists of many participants with the ability to affect and influence each other. They derive this concept from the observation that people seek richness, complexity, and depth within a virtual world.

VIRTUAL ENVIRONMENTS

Although the present-day, loose use of the term "cyberspace" no longer implies or suggests immersion in a virtual reality, current technology allows the integration of a number of capabilities (sensors, signals, connections, transmissions, processors, and controllers) sufficient to generate a virtual interactive experience that is accessible regardless of a geographic location.

RECENT DEFINITIONS OF CYBERSPACE

The most recent draft definition is the following (Strate, 1999): Cyberspace is a global and dynamic domain (subject to constant change) characterized by the combined use of electrons and electromagnetic spectrum, whose purpose is to create, store, modify, exchange, share and extract, use, and eliminate information and disrupt physical resources.

Cyberspace includes the following:

a. Physical infrastructures and telecommunications devices that allow for the connection of technological and communication system networks, understood in the broadest sense (Supervisory Control and Acquisition Data devices, smartphones/tablets, computers, servers, etc.)
b. Computer systems (see point a) and the related (sometimes embedded) software that guarantee the domain's basic operational functioning and connectivity
c. Networks between computer systems
d. Networks of networks that connect computer systems (the distinction between networks and networks of networks is mainly organizational)
e. The access nodes of users and intermediaries routing nodes
f. Constituent data (or resident data). Often, in common parlance (and sometimes in commercial language), networks of networks are called internet (with a lowercase i), while networks between computers are called intranet. Internet (with a capital I, in journalistic language sometimes called the Net) can be considered a part of system a)

A distinctive and constitutive feature of cyberspace is that no central entity exercises control over all the networks that make up this new domain.

CYBER THREAT INTELLIGENCE

Cyber threat intelligence (CTI) is an "elusive" concept. While cybersecurity comprises the recruitment of IT security experts and the deployment of technical means, to protect an organization's critical infrastructure or intellectual property, CTI is based on the collection of intelligence using open-source intelligence (OSINT), social media intelligence, human intelligence (HUMINT), or intelligence from the deep and dark web. CTI's key mission is to research and analyze trends and technical developments in three areas:

- Cyber crime
- Cyber activism
- Cyber espionage (advanced persistent threat or APT)

Those accumulated data based on research and analysis enable states to come up with preventive measures in advance. Considering the seriousness of the impacts of cyber threats, CTI has been raised as an efficient solution to maintain international security.

Types of CTI

According to UK's Centre for the Protection of National Infrastructure (CPNI), there are four types of threat intelligence (CPNI, 2017):

1. Tactical: Attacker methodologies, tools, and tactics; relies on enough resources and involves certain actions to go against potentially dangerous actors trying to do infiltration.
2. Technical: Indicators of specific malware.
3. Operational: Details of specific incoming attack; assess the organization's ability in determining the future cyber threats.
4. Strategic: High-level information on changing risk (strategic shifts); senior leadership is required for thorough determination to critically assess threats.

In the financial sector, the CBEST* framework of the Bank of England assumes that penetration testing is no longer adequate to protect sensitive business sectors, such as the banking sector. In response, the UK Financial Authorities (Bank of England, Her Majesty's Treasury, and the Financial Conduct Authority) recommend several steps to guard financial institutions from cyber threats, including receiving "advice from the CTI providers operating within the UK Government" (Bank of England, 2016).

Benefits of Tactical Cyber Intelligence

- Provides context and relevance to a tremendous amount of data.
- Empowers organizations to develop a proactive cybersecurity posture and bolster its overall risk management policies.
- Informs better decision making during and following the detection of a cyber-intrusion.
- Drives momentum toward a cybersecurity posture that is predictive, not just reactive.

THE CHALLENGE OF ATTRIBUTION

Behind any cyber threat, there are people using computers and networks. During or after a cyber-attack, technical information about the network and computers between the attacker and the victim can be collected. However, identifying the person(s) behind an attack, their motivations, or the ultimate sponsor of the attack is difficult. Recent efforts in threat intelligence emphasize understanding adversary tactics, techniques, and procedures (TTPs).

* To assist the boards of financial firms and infrastructure providers, and regulators, in improving their understanding of the types of cyber-attack that could undermine financial stability in the UK, and the extent to which the UK financial sector is vulnerable to those attacks, a new, intelligence-led testing framework has been devised by the UK Financial Authorities in conjunction with CREST (the Council for Registered Ethical Security Testers) and Digital Shadows.

CYBER WARFARE

Cyber warfare has been defined as "actions by a nation-state to penetrate another nation's computers or networks for the purposes of causing damage or disruption" (IMSM, 2010), but other definitions also include nonstate actors, such as terrorist groups, companies, political or ideological extremist groups, hacktivists, and transnational criminal organizations.

Some governments have made it an integral part of their overall military strategy, with some having invested heavily in cyber warfare capability. Cyber warfare is essentially a formalized version of penetration testing in which a government entity has established it as a warfighting capability. This capability uses the same set of penetration testing methodologies but applies them, in the case of U.S. doctrine, in a strategic way to

- Prevent cyber-attacks against America's critical infrastructure
- Reduce national vulnerability to cyber-attacks
- Minimize damage and recovery time from cyber-attacks

Offensive operations are also part of these national level strategies for officially declared wars as well as undeclared secretive operations.

Types of Threat

1. Cyber-attacks, where damage or disruption is caused are the main concern
2. Cyber espionage, which can provide the information needed to launch a successful attack

Espionage

Traditional espionage is not an act of war, nor is cyber-espionage, and both are generally assumed to be ongoing between major powers.

Despite this assumption, some incidents can cause serious tensions between nations and are often described as "attacks." Examples are the following:

- Massive spying by the United States on many countries, revealed by Edward Snowden.
- After the National Security Administration's (NSA's) spying on Germany's Chancellor Angela Merkel was revealed, the Chancellor compared the NSA with the Stasi (Traynor, 2013).
- NSA recording nearly every cell phone conversation in the Bahamas without the Bahamian government's permission, and similar programmers in Kenya, the Philippines, Mexico, and Afghanistan.
- The "Titan Rain" probes of American defense contractor computer systems since 2003.
- The Office of Personnel Management data breach, in the United States, widely attributed to China.

Sabotage

Computers and satellites that coordinate other activities are vulnerable components of a system and could lead to the disruption of equipment. Compromise of military systems, such as C4ISTAR components that are responsible for orders and communications, could lead to their interception or malicious replacement. Power, water, fuel, communications, and transportation infrastructure all may be vulnerable to disruption. Especially the civilian realm is also at risk, noting that the security breaches have already gone beyond stolen credit card numbers and that potential targets can also include the electric power grid, trains, or the stock market.

In mid-July 2010, security experts discovered a malicious software program called Stuxnet that had infiltrated factory computers and had spread to plants around the world. It is considered "the first attack on critical industrial infrastructure that sits at the foundation of modern economies," notes *The New York Times* (Zetter, 2014).

Stuxnet, while extremely effective in delaying Iran's nuclear program for the development of nuclear weaponry, came at a high cost. For the first time, it became clear that not only could cyber weapons be defensive but they could also be offensive. The large decentralization and scale of cyberspace makes it extremely difficult to direct from a policy perspective.

Nonstate actors can play as large a part in the cyberwar space as state actors can, which leads to dangerous, sometimes disastrous, consequences.

Small groups of highly skilled malware developers are able to as effectively impact global politics and cyber warfare as large governmental agencies. A major aspect of this ability lies in the willingness of these groups to share their exploits and developments on the web as a form of arms proliferation.

This allows lesser hackers to become more proficient in creating the large-scale attacks that once only a small handful were skillful enough to manage. In addition, thriving black markets for these kinds of cyber weapons are buying and selling these cyber capabilities to the highest bidder without regard for consequences.

DENIAL-OF-SERVICE ATTACK

In computing, a denial-of-service attack (DoS attack) or distributed DoS attack is an attempt to make a machine or network resource unavailable to its intended users. Perpetrators of DoS attacks typically target sites or services hosted on high-profile web servers such as banks, credit card payment gateways, and even root name servers.

DoS attacks may not be limited to computer-based methods, as strategic physical attacks against infrastructure can be just as devastating. For example, cutting undersea communication cables may severely cripple some regions and countries with regard to their information warfare ability.

ELECTRICAL POWER GRID

The federal government of the United States admits that the electric power grid is susceptible to cyber warfare. The U.S. Department of Homeland Security works with industries to identify vulnerabilities and to help industries enhance the security of control system networks, the federal government is also working to ensure that security is built in as the next generation of "smart grid" networks is developed.

In April 2009, reports surfaced that China and Russia had infiltrated the U.S. electrical grid and left behind software programs that could be used to disrupt the system, according to current and former national security officials. The North American Electric Reliability Corporation (NERC) has issued a public notice that warns that the electrical grid is not adequately protected from cyber-attack.

China denies intruding into the U.S. electrical grid. One countermeasure would be to disconnect the power grid from the Internet and run the net with droop speed control only. Massive power outages caused by a cyber-attack could disrupt the economy, distract from a simultaneous military attack, or create a national trauma.

MOTIVATIONS

Military

In the United States, General Keith B. Alexander, first head of the recently formed USCYBERCOM, told the Senate Armed Services Committee that computer network warfare is evolving so rapidly that there is a "mismatch between our technical capabilities to conduct operations and the governing laws and policies. Cyber Command is the newest global combatant and its sole mission is cyberspace, outside the traditional battlefields of land, sea, air and space" (Sensagent, 2016). It will attempt to find and, when necessary, neutralize cyber-attacks and to defend military computer networks.

Now, the issue for the computer warfare command is envisioned, listing the kind of targets that his or her new headquarters could be ordered to attack, including "traditional battlefield prizes—command-and-control systems at military headquarters, air defense networks and weapons systems that require computers to operate."

One cyber warfare scenario, Cyber Shockwave, which was war gamed on the cabinet level by former administration officials, raised issues ranging from the National Guard to the power grid to the limits of statutory authority.

The distributed nature of internet-based attacks means that it is difficult to determine motivation and attacking party, meaning that it is unclear when a specific act should be considered an act of war.

Civil

Potential targets in internet sabotage include all aspects of the Internet from the backbones of the web, to the internet service providers, to the varying types of data communication mediums and network equipment. This would include web servers, enterprise information systems, client server systems, communication links, network equipment, and the desktops and laptops in businesses and homes. Electrical grids and telecommunication systems are also deemed vulnerable, especially due to current trends in automation.

Hacktivism

Politically motivated hacktivism involves the subversive use of computers and computer networks to promote an agenda and can potentially extend to attacks, theft, and virtual sabotage that could be seen as cyber warfare—or mistaken for it.

Private Sector

Computer hacking represents a modern threat in ongoing industrial espionage and as such is presumed to widely occur. It is typical that this type of crime is underreported. Corporations around the world face millions of cyber-attacks a day. Most of these attacks don't gain any media attention or lead to strong political statements by victims. This type of crime is usually financially motivated.

CYBER SPYING

Cyber spying, or cyber espionage, is the act or practice of obtaining secrets without the permission of the holder of the information (personal, sensitive, proprietary, or of classified nature), from individuals, competitors, rivals, groups, governments, and enemies for personal, economic, political, or military advantage using methods on the Internet, networks, or individual computers through the use of cracking techniques and malicious software including Trojan horses and spyware.

It may wholly be perpetrated online from computer desks of professionals on bases in faraway countries or may involve infiltration at home by computer-trained conventional spies and moles or in other cases may be the criminal handiwork of amateur malicious hackers and software programmers.

Details

Cyber spying typically involves the use of such access to secrets and classified information or control of individual computers or whole networks for a strategic advantage and for psychological, political, and physical subversion activities and sabotage. More recently, cyber spying involves analysis of public activity on social networking sites like Facebook and Twitter.

Such operations, like non-cyber-espionage, are typically illegal in the victim country while fully supported by the highest level of government in the aggressor country.

The ethical situation likewise depends on one's viewpoint, particularly one's opinion of the governments involved.

In response to reports of cyber spying by China against the United States, Amitai Etzioni of the Institute for Communitarian Policy Studies has suggested that China and the United States should agree to a policy of mutually assured restraint with respect to cyberspace (Revolvy, 2017).

This would involve allowing both states to take the measures they deem necessary for their self-defense while simultaneously agreeing to refrain from taking offensive steps or engaging in cyber espionage; it would also entail vetting these commitments. In September 2015, the United States and China agreed not to allow parties in their nations to cyber spy on each other for commercial gain, but did not prohibit government spying.

Proactive Cyber Defense

Proactive cyber defense or active cyber defense means acting in anticipation to oppose an attack against computers and networks. Proactive cyber defense will most often require additional security from internet service providers.

Some of the reasons for a proactive defense strategy are about cost and choice. Making choices after an attack is difficult and costly. Proactive defense is key to mitigating operational risk.

Cyberspace typically refers to the vast and growing logical domain composed of public and private networks; independently managed networks linked together through the lingua franca of the Internet, the Internet Protocol (IP).

The definition of cyberspace has been extended to include all network-space, which at some point, through some path, may have eventual access to the public internet. Under this definition, cyberspace becomes virtually every networked device in the world, which is not devoid of a network interface entirely. There is no air-gap anymore between networks.

The origins of cyber defense undoubtedly evolved from the original purpose of the Internet, which was to harden military networks against the threat of a nuclear strike. Later cyber defense was coveted by the tenets of information warfare and information operations.

Current Status

Information warfare is an emergent reality that comes from a self-organization process that has never been seen before. The problem is that we talk about it using terms that have well-known connotations. And it is difficult to talk about something completely new using words that bring with them specific understanding and expectancies. The early period of the automobile faced a similar situation. At one time, it was called a "horseless carriage," as this was the only way to define its essential quality.

The car is more than a carriage without a horse. This is the dilemma we face when we discuss information warfare. The danger is that the uses of familiar words misrepresent and mask the true extend of the revolution that will have to take place if we are to be able to retain a military capacity in a new physical, social, and cognitive space.

Proactive Preemptive Operations

Effective cyber defenses ideally prevent an incident from taking place. Any other approach is simply reactive. Federal Technology Service Federal Computer Incident Response Center, the National Infrastructure Protection Center, the National Structural Integrity Research Centre, the Department of Defense (DoD), and industry components realize that the best [action] is a preemptive and proactive approach.

The mission of the Proactive, Preemptive Operations Group (P2OG; a U.S. intelligence agency that would employ "black world" [black operations] tactics) is reportedly to conduct aggressive, proactive, and preemptive operations to interdict and disrupt the threat using psychological operations, managed information dissemination, precision targeting, information warfare operations, and signals intelligence (SIGINT). The proactive defense strategy is meant to improve information collection by stimulating reactions of the threat agents, provide strike options, and enhance operational preparation of the real or virtual battle space.

The P2OG has been recommended to be constituted of "one hundred 'highly specialized' people with unique technical and intelligence skills such as information operations, PSYOPS, network attack, covert activities, SIGINT, HUMINT, SOF, influence warfare/deception operations and to report to the National Security Council."

The DoD doctrinally would initiate a "preemptive" attack on the basis of evidence that an enemy attack is imminent. Proactive measures, according to the DoD, are those actions taken directly against the preventive stage of an attack by the enemy.

Strike-back doctrine aligns with the preemptive and counterattack tactics of a proactive cyber defense strategy.

NATIONAL STRATEGY TO SECURE CYBERSPACE

In the U.S. government, the National Strategy to Secure Cyberspace is a component of the larger National Strategy for Homeland Security. The National Strategy to Secure Cyberspace was drafted by the Department of Homeland Security in reaction to the September 11, 2001, terrorist attacks (Department of Homeland Security, 2014).

Released on February 14, 2003, it offers suggestions, not mandates, to business, academic, and individual users of cyberspace to secure computer systems and networks. It was prepared after a year of research by businesses, universities, and government and after five months of public comment. The plan advises a number of security practices as well as promotion of cybersecurity education.

The National Strategy to Secure Cyberspace identifies three strategic objectives:

1. Prevent cyber-attacks against America's critical infrastructures
2. Reduce national vulnerability to cyber-attacks
3. Minimize damage and recovery time from cyber-attacks that do occur

To meet these objectives, the National Strategy outlines five national priorities:

- The first priority, the creation of a National Cyberspace Security Response System, focuses on improving the government's response to cyberspace security incidents and reducing the potential damage from such events.
- The second, third, and fourth priorities (the development of a National Cyberspace Security Threat and Vulnerability Reduction Program, the creation of a National Cyberspace Security Awareness and Training Program, and the necessity of Securing Governments' Cyberspace) aim to reduce threats from, and vulnerabilities to, cyber-attacks.

- The fifth priority, the establishment of a system of National Security and International Cyberspace Security Cooperation, intends to prevent cyber-attacks that could impact national security assets and to improve the international management of and response to such attacks.

Ultimately, the National Strategy encourages companies to regularly review their technology security plans and individuals who use the Internet to add firewalls and antivirus software to their systems. It calls for a single federal center to help detect, monitor, and analyze attacks and for expanded cybersecurity research and improved government–industry cooperation.

CYBER-HUMINT

Cyber-Humint refers to the set of skills used by hackers, within cyberspace, in order to obtain private information while attacking the human factor, using various psychological deceptions. Cyber-Humint includes the use of traditional human espionage methodologies, such as agent recruitment, information gathering through deception (traditionally known as Humint), combined with deception technologies known as social engineering.

Background in Cyber-Humint

Intelligence gathering involves a range of specialized approaches—from SIGINT, imagery intelligence, measurement and signature intelligence, and geospatial intelligence, to OSINT. In many cases, information collected from human sources is still considered highly reliable by intelligence analysts, especially while transforming a collection of disparate data strands into an actionable prevention plan.

Cyber-Humint methodology was first coined by Ed Alcantara AFX DBI in February 2010 (AOL Inc., 2017). Amit Steinhart argued that the cooperation between skilled Humint experts trained with specific Humint capabilities and computer security specialists, who apply "social engineering" techniques, is one of the main advantages of Cyber-Humint. Steinhart offered a new model of information security strategy that imports concepts from Humint espionage and combines it with social engineering strategies, such as the usage of avatars for agents operating in cyberspace or information and disinformation spreading through cyberspace.

Humint experts often argue that in comparison to the relatively young social engineering concept, Humint practices, which had been developed for many years by professionals working at national intelligence services, hold the higher ground in terms of experience, technologies, and practices. A new form of cyber capability was created when the technical capabilities of computer experts were combined with the intelligence experience of Humint experts.

Cyber-Humint Strategy Orientation

Cyber-Humint is aimed to effectively defend organizations against APT attacks. In the beginning of the 2010s, organizations such as the American NSA and British GCHQ have started to invest significant resources into acquiring technological and intelligence capabilities, to help identify cyber aggressors and assess their abilities and tactical skills.

Recently, information security has shifted from building firewalls to building systems, in order to provide real-time intelligence. Most near-future scenarios suggest that organizations who fail to adapt to the systematic cyber approach will find themselves in a critical situation.

In 2011, attention was drawn to the fact that while cybersecurity experts can deliver extensive reports on Internet risks, most of the alerts are still general and unspecific and do not actually meet the expectations of the specific organization. In addition, cybersecurity companies locate hackers or cyber-attackers only when the attack is already in progress or, worse, after a given system has already been damaged or compromised.

The majority of cybersecurity defenders currently use automatic network scans as a routine measure. A human analyst becomes involved only at the final stage of data gathering, which means that the bulk of the available data will not be analyzed in real time.

Hackers and Cyber-Humint

The majority of cybersecurity companies have no access to human operators within the dark web. Hence, they do not benefit from the key input of informants and agent provocateurs. These companies do not apply the methods of agent recruitment and agent management, which various national intelligence organizations have developed and used effectively for years.

New information technologies allow hackers to acquire the upper hand in any confrontation with the targeted organization. A case in point is APT, which in impact and devastation equals to a military strike against a civilian entity. Many peripheral defense systems are not capable of recognizing indications of incoming attacks in advance and cannot intercept the attack during its course. The majority of security systems can acknowledge the attack only after the damage has already occurred.

Most organizations prefer to focus their security efforts on inward-facing protection strategies, in an attempt to prevent attackers from entering the organization's network. Their defense protocols are not designed to protect from attempts to exploit the organization's employees, who have become the main target for willful intelligence gathering.

Personal behavior, compromising private situations, work habits, passwords, and other private and business information can be easily harvested and used to facilitate an attack against the organization.

The Interface Between Cyber Experts and Cyber-Humint

The concept of Cyber-Humint allows cyber experts and Humint specialists to use real-life human sources, both in and within many public or secret online social networks and operating systems.

By investigating authentic human sources, intelligence experts and cyber experts can explore the various possible aims of potential attackers and their abilities, by monitoring their electronic activities. Outcomes usually leave much to be desired. Attackers are identified only after the attack has started. In just a handful of cases did companies manage to alert their clients against a pending attack.

Cyber-Humint involves recruiting human agents and deploying them with strategic efficiency to provide the organization with a clear, focused picture of likely threats and hostile actors with the intention of harming the organization. Cyber-Humint uses classic Humint tactics that had been practiced for more than half a century by the national intelligence agencies. It combines them with hackers' social engineering concepts.

Using Cyber-Humint requires qualified computer professionals who are well versed in the behavior patterns, linguistic nuances, and conventions accepted within the Darknet, as well as other online networks and subcultures.

Conversant computer experts and intelligence specialists work in synchrony to uncover indications of intent, long before it develops into an attack plan, so organizations can decide how, where, and when to expose or incapacitate the potential attackers.

CYBERSECURITY STANDARDS

Cybersecurity standards (also styled cyber security standards) are techniques generally set forth in published materials that attempt to protect the cyber environment of a user or organization. This environment includes users themselves, networks, devices, all software, processes, information in storage or transit, applications, services, and systems that can be connected directly or indirectly to networks.

Principal objective is to reduce the risks, including prevention or mitigation of cyber-attacks. These published materials consist of collections of tools, policies, security concepts, security safeguards, guidelines, risk management approaches, actions, training, best practices, assurance, and technologies.

EUROPEAN TELECOMMUNICATIONS STANDARDS INSTITUTE (ETSI) CYBERSECURITY TECHNICAL COMMITTEE

The ETSI Cybersecurity Technical Committee (TC CYBER) is responsible for the standardization of cybersecurity internationally and for providing a center of relevant expertise for other ETSI committees. Growing dependence on networked digital systems has brought with it an increase in both the variety and quantity of cyber-threats.

The different methods governing secure transactions in the various Member States of the European Union sometimes make it difficult to assess the respective risks and to ensure adequate security. Building on ETSI's world-leading expertise in the security of information and communications technologies, it set up a new cybersecurity committee (TC CYBER) in 2014 to meet the growing demand for standards to protect the Internet and the communications and business it carries.

TC CYBER is working closely with relevant stakeholders to develop appropriate standards to increase privacy and security for organizations and citizens across Europe.

The committee is looking in particular at the security of infrastructures, devices, services, and protocols, as well as security tools and techniques to ensure security. It offers security advice and guidance to users, manufacturers, and network and infrastructure operators. Its standards are freely available online. A principal work item effort is the production of a global cybersecurity ecosystem of standardization and other activities.

ISO 27001 AND 27002

ISO/IEC 27001:2013, part of the growing ISO/IEC 27000 family of standards, is an information security management system (ISMS) standard published in October 2013 by the International Organization for Standardization (ISO) and the International Electro technical Commission (IEC) (Consultant.com, 2017). Its full name is ISO/IEC 27001:2013—Information technology—Security techniques—Information security management systems—Requirements.

ISO/IEC 27001:2013 formally specifies a management system that is intended to bring information security under explicit management control.

ISO/IEC 27002 incorporates mainly part 1 of the BS 7799 good security management practice standard. The latest version of BS7799 is BS7799-3. Sometimes, ISO/IEC 27002 is therefore referred to as ISO 17799 or BS 7799 part 1, and sometimes, it refers to part 1 and part 7. BS 7799 part 1 provides an outline or good practice guide for cybersecurity management, whereas BS 7799 part 2 and ISO 27001 are normative and therefore provide a framework for certification. ISO/IEC 27002 is a high-level guide to cybersecurity.

It is most beneficial as an explanatory guide for the management of an organization to obtain certification to the ISO 27001 standard. The certification, once obtained, lasts three years. Depending on the auditing organization, no or some intermediate audits may be carried out during the three years.

ISO 27001 (ISMS) replaces BS 7799 part 2, but since it is backward compatible, any organization working toward BS 7799 part 2 can easily transition to the ISO 27001 certification process. There is also a transitional audit available to make it easier once an organization is BS 7799 part 2 certified for the organization to become ISO 27001 certified. ISO/IEC 27002 provides best practice recommendations on information security management for use by those responsible for initiating, implementing, or maintaining ISMS. It states the information security systems required to implement ISO 27002 control objectives. Without ISO 27001, ISO 27002 control objectives are ineffective. ISO 27002 controls objectives are incorporated into ISO 27001 in Annex A.

ISO/IEC 21827 (SSE-CMM–ISO/IEC 21827) is an international standard based on the Systems Security Engineering Capability Maturity Model (SSE-CMM) that can measure the maturity of ISO control objectives.

STANDARD OF GOOD PRACTICE

In the 1990s, the Information Security Forum (ISF) published a comprehensive list of best practices for information security, published as the Standard of Good Practice (SoGP) (ISF, 2017). The ISF continues to update the SoGP every two years (with the exception of 2013–2014); the latest version was published in 2016.

Originally, the SoGP was a private document available only to ISF members, but the ISF has since made the full document available for sale to the general public.

Among other programs, the ISF offers its member organizations a comprehensive benchmarking program based on the SoGP. Furthermore, it is important for those in charge of security management to understand and adhere to NERC CIP compliance requirements.

NATIONAL INSTITUTE OF STANDARDS AND TECHNOLOGY

1. The National Institute of Standards and Technology (NIST) Cybersecurity Framework provides a high level taxonomy of cybersecurity outcomes and a methodology to assess and manage those outcomes. It is intended to help private sector organizations that provide critical infrastructure with guidance on how to protect it, along with relevant protections for privacy and civil liberties.
2. Special publication 800-12 provides a broad overview of computer security and control areas. It also emphasizes the importance of the security controls and ways to implement them. Initially, this document was aimed at the federal government, although most practices in this document can be applied to the private sector as well. Specifically, it was written for those people in the federal government responsible for handling sensitive systems.
3. Special publication 800-14 describes common security principles that are used. It provides a high-level description of what should be incorporated within a computer security policy. It describes what can be done to improve existing security as well as how to develop a new security practice.
4. Eight principles and 14 practices are described within this document.
5. Special publication 800-26 provides advice on how to manage IT security. It is superseded by NIST SP 800-53 rev3. This document emphasizes the importance of self-assessments as well as risk assessments.
6. Special publication 800-37, updated in 2010, provides a new risk approach: "Guide for Applying the Risk Management Framework to Federal Information Systems."
7. Special publication 800-53 rev4, "Security and Privacy Controls for Federal Information Systems and Organizations," was published in April 2013 and updated to include updates as of January 15, 2014. It specifically addresses the 194 security controls that are applied to a system to make it "more secure."

8. Special Publication 800-82, Revision 2, "Guide to Industrial Control System (ICS) Security," revised in May 2015, describes how to secure multiple types of Industrial control systems against cyber-attacks while considering the performance, reliability, and safety requirements specific to ICS.

In summary, the principal objective is to reduce the risks, including prevention or mitigation, of cyber-attacks. These published materials consist of collections of tools, policies, security concepts, security safeguards, guidelines, risk management approaches, actions, training, best practices, assurance, and technologies.

15 Emergency and the National Incident Management System (NIMS)

Emergency management and the National Incident Management System (NIMS) go hand in hand; one cannot exist without the other. It's a great tool to plan and write one's Local Emergency Response Plan (LERP) and perform the analysis process.

The NIMS provides a systematic, proactive approach to guide departments and agencies at all levels of government, nongovernmental organizations (NGOs), and the private sector to work seamlessly to prevent, protect against, respond to, recover from, and mitigate the effects of incidents, regardless of cause, size, location, or complexity, in order to reduce the loss of life and property and harm to the environment.

NIMS works hand in hand with the National Response Framework (NRF). NIMS provides the template for the management of incidents, while the NRF provides the structure and mechanisms for national-level policy for incident management.

INTRODUCTION

Homeland Security Presidential Directive 5—Management of Domestic Incidents called for the establishment of a single, comprehensive national incident management system.

As a result, the U.S. Department of Homeland Security released the NIMS in March (Department of Homeland Security, 2008). The NIMS provides a systematic, proactive approach guiding departments and agencies at all levels of government, the private sector, and NGOs to work seamlessly to prepare for, prevent, respond to, recover from, and mitigate the effects of incidents, regardless of cause, size, location, or complexity, in order to reduce the loss of life, property, and harm to the environment.

This consistency provides the foundation for implementation of the NIMS for all incidents, ranging from daily occurrences to incidents requiring a coordinated Federal response. The NIMS document, as revised in 2007, reflects contributions from stakeholders and lessons learned during recent incidents.

States and territories play an important role in ensuring effective NIMS implementation; not only must they implement NIMS within state and territory departments and agencies but also ensure that the systems and processes are in place to communicate and support NIMS compliance at all jurisdictional levels. In some instances, when tribal nations or local government may not have the resources to implement all NIMS elements on their own, states and territories should support their efforts by encouraging them to cooperate with other localities in their regions and pool their resources to implement NIMS.

The long-term goal of NIMS is to provide a consistent framework for all aspects of emergency management and incident response. This framework should be sustainable, flexible, and scalable to meet changing incident needs and allow for integration of other resources from various partners through mutual aid agreements and/or assistance agreements.

The following are the NIMS Compliance Objectives and a path to follow when performing emergency management planning process and setting up objectives.

I. What needs to happen?
   1. Adopt NIMS for all Departments/Agencies, as well as promote and encourage NIMS adoption by associations, utilities, nongovernmental organizations (NGOs) and private sector emergency management and incident response organizations.
   2. Monitor formal adoption of NIMS, to include formal NIMS adoption by Departments/Agencies.
   3. Establish and maintain a planning process to communicate, monitor and implement all NIMS requirements across the State/Territory (including Departments/Agencies), including local governments. This process must provide a means for measuring progress and facilitate reporting.
   4. Designate and maintain a single point of contact within government to serve as principal coordinator for NIMS implementation jurisdiction-wide (to include a principal coordinator for NIMS implementation within each Department/Agency).
   5. Ensure that Federal Preparedness Awards [to include, but not limited to, DHS Homeland Security Grant Program and Urban Area Security Initiative Funds] to State/Territorial Departments/Agencies, as well as local governments, support all required NIMS compliance Objectives.
   6. Audit agencies and review organizations should routinely include NIMS compliance requirements in all audits associated with Federal Preparedness Awards.
   7. Assist Tribal Nations with formal adoption and implementation of NIMS.

II. Preparedness: Planning
   1. Revise and update emergency operations plans (EOPs), standard operating procedures (SOPs), and standard operating guidelines (SOGs) to incorporate NIMS and NRF components, principles and policies, to include planning, training, response, exercises, equipment, evaluation, and corrective actions.
   2. Promote and/or develop intrastate and interagency mutual aid agreements and assistance agreements (to include agreements with the private sector and NGOs).

3. Include preparedness organizations and elected and appointed officials in the development of emergency operations plans (EOPs).
4. Plan for special needs populations in the development of EOPs (to include, but not limited to, individuals with limited English language proficiency; individuals with disabilities; children; the aged, etc.).
5. Use existing resources such as programs, personnel, and training facilities to coordinate and deliver NIMS training requirements.
6. Promote and encourage delivery of NIMS training (as identified in the Five-Year NIMS Training Plan Schedule, December 2007).

III. Preparedness: Exercise
1. Incorporate NIMS concepts and principles into all appropriate State/Territorial training and exercises.
2. Plan for and/or participate in an all-hazards exercise program [for example, Homeland Security Exercise and Evaluation Program] that involves emergency management/response personnel from multiple disciplines and/or multiple jurisdictions.
3. Incorporate corrective actions into preparedness and response plans and procedures.
4. Include NGOs and the private sector in an all-hazards exercise program, when appropriate.
5. Promote the integration of Incident Command, MACS, and Public Information into appropriate exercises.

IV. Communications and Information Management
1. Apply common and consistent terminology as used in NIMS, including the establishment of plain language (clear text) communications standards.
2. Utilize systems, tools, and processes to present consistent and accurate information (e.g., common operating picture) during an incident/planned event.
3. Institute procedures and protocols for operational and information security during an incident/planned event.
4. Institute multidisciplinary and/or multi-jurisdictional procedures and protocols for standardization of data collection and analysis to utilize or share information during an incident/planned event.
5. Develop procedures and protocols for communications (to include voice, data, access to geospatial information, Internet/Web use, and data encryption), where applicable, to utilize or share information during an incident/planned event.

V. Resource Management
1. Inventory response assets to conform to NIMS National Resource Typing.
2. Definitions, as defined by FEMA Incident Management Systems Division.
3. Ensure that equipment, communications and data systems acquired through State/Territorial and local acquisition programs are interoperable.

4. Utilize response asset inventory for intrastate and interstate mutual aid requests [such as Emergency Management Assistance Compact (EMAC)], training, exercises, and incidents/planned events.
5. Initiate State/Territory-wide system to credential emergency management/response personnel to ensure proper authorization and access to an incident including those involving mutual aid agreements and/or assistance agreements.
6. Inventory and type specific emergency management/response resources and assets to address unique needs beyond current "Tier One" NIMS National Resource Typing Definitions.
7. Institute policies, plans, procedures and protocols to prevent spontaneous deployment of resources/personnel and/or responding to a request that bypassed official resource coordination processes (i.e., resources requested through improper channels).
8. Institute mechanisms to deploy, track, recover, demobilize, and to provide reimbursement for resources utilized during response and recovery.

VI. Command and Management

1. Incident Command System (ICS): Manage all incidents/planned events in accordance with ICS organizational structures, doctrine and procedures.
2. ICS implementation must include the consistent application of Incident Action Planning (IAP), common communications plans, implementation of Area Command (AC) to oversee multiple incidents that are handled by separate ICS organizations or to oversee the management of a very large or evolving incident that has multiple incident management teams engaged, and implementation of unified command (UC) in multijurisdictional or multiagency incident management, as appropriate.
3. Multi-agency Coordination System (MACS): Coordinate and support emergency management and incident response Objectives through the development and use of integrated multi-agency coordination systems, i.e.,—develop and maintain connectivity capability between local Incident Command Posts (ICP), local 911 Centers, local Emergency Operations Centers (EOCs), the State/Territorial EOC and Regional and/Federal EOCs and NRF organizational elements.
4. Public Information: Institutionalize, within the framework of ICS, Public Information, [e.g., Joint Information System (JIS) and a Joint Information Center (JIC)] during an incident/planned event.
5. Ensure that Public Information procedures and processes can gather, verify, coordinate, and disseminate information during an incident/planned event.
6. Utilize access control measures during an incident, as appropriate.

In summary, the long-term goal of NIMS is to provide a consistent framework for all aspects of emergency management and incident response. This framework should be sustainable, flexible, and scalable to meet changing incident needs and allow for integration of other resources from various partners through mutual aid agreements and/or assistance agreements. Using this format and guidelines from your community surveys and hazard analysis and vulnerability surveys will help one create and maintain a common sense approach to the emergency planning process.

It is recommended for all response personnel and emergency management personnel to have the following courses from Federal Emergency Management Agency (FEMA). These courses are online and of no charge to the agency and local government.

ICS and NIMS courses:

- ICS-100: Introduction to the Incident Command System
- ICS-200: ICS for Single Resources and Initial Action Incidents
- ICS-300: Intermediate ICS for Expanding Incidents
- ICS-400: Advanced ICS for Command and General Staff
- IS-700: National Incident Management System, An Introduction
- IS-701: NIMS Multiagency Coordination System (MACS)
- IS-702: NIMS Publication Information Systems
- IS-703: NIMS Resource Management
- IS-704: NIMS Communication and Information Management (unavailable)
- IS-706: NIMS Intrastate Mutual Aid—An Introduction
- IS-800: National Response Framework, An Introduction
- G-191: Incident Command System/Emergency Operations Center Interface
- G-402: Incident Command System (ICS) Overview for Executives/Senior Officials
- G-775: Emergency Operations Center (EOC) Management and Operations

NIMS courses can be located at https://training.fema.gov/nims/.

ICS 300/400, G191, G402, and G-775 are coordinated by local Emergency Management Agencies, please contact them directly for course offerings in your area. Contact information for State or Territorial Emergency Management Agencies can be found at https://training.fema.gov/programs/aps/stolist.aspx.

All-Hazards Position Specific courses:

- E/L 950: All-Hazards Position Specific Incident Commander
- E/L 952: All-Hazards Position Specific Public Information Officer
- E/L 954: All-Hazards Position Specific Safety Officer
- E/L 956: All-Hazards Position Specific Liaison Officer
- E/L 958: All-Hazards Position Specific Operations Section Chief
- E/L 960: All-Hazards Position Specific Division/Group Supervisor
- E/L 962: All-Hazards Position Specific Planning Section Chief
- E/L 964: All-Hazards Position Specific Situation Unit Leader
- E/L 965: All-Hazards Position Specific Resources Unit Leader
- E/L 967: All-Hazards Position Specific Logistics Section Chief
- E/L 969: All-Hazards Position Specific Communications Unit Leader
- E/L 970: All-Hazards Position Specific Supply Unit Leader
- E/L 971: All-Hazards Position Specific Facilities Unit Leader
- E/L 973: All-Hazards Position Specific Finance/Admin. Section Chief
- E/L 975: All-Hazards Position Specific Finance/Admin. Unit Leader Course
- E/L 984: Task Force/Strike Team Leader

- E/L 986: Air Support Group Supervisor
- E/L 987: Introduction to Air Operations

The NIMS ICS All-Hazards Position Specific Training Program official website is at https://training.fema.gov/allhazards/.

SUMMARY

The following are emergency management skills, training, and experiences needed to be an effective emergency manager and direct these operations.

- Coordinate disaster response or crisis management activities, such as ordering evacuations, opening public shelters, and implementing special needs plans and programs.
- Prepare plans that outline operating procedures to be used in response to disasters or emergencies, such as hurricanes, nuclear accidents, and terrorist attacks, and in recovery from these events.
- Develop and maintain liaisons with municipalities, county departments, and similar entities to facilitate plan development, response effort coordination, and exchanges of personnel and equipment.
- Design and administer emergency or disaster preparedness training courses that teach people how to effectively respond to major emergencies and disasters.
- Keep informed of activities or changes that could affect the likelihood of an emergency, as well as those that could affect response efforts and details of plan implementation.

Technology used:

- Database user interface and query software—Emergency Managers Weather Information Network, Federal Emergency Management Information System, relational database software, and SoftRisk Technologies (SoftRisk SQL)
- Electronic mail software—IBM Lotus Notes
- Map creation software—Digital Engineering Corporation E-MAPS; ESRI ArcGIS software Hot technology; Geographic information system (GIS) software Hot technology; MapInfo Professional
- Project management software—Alert Technologies OpsCenter; Emergency Services Integrators ESi WebEOC; Microsoft SharePoint software Hot technology; National Center for Crisis and Continuity Coordination NC4 E Team
- Spreadsheet software—Microsoft Excel Hot technology

Leadership responsibilities to ensure compliance and readiness

- *Public safety and security*—Knowledge of relevant equipment, policies, procedures, and strategies to promote effective local, state, or national security operations for the protection of people, data, property, and institutions.

- *Law and government*—Knowledge of laws, legal codes, court procedures, precedents, government regulations, executive orders, agency rules, and the democratic political process.
- *Administration and management*—Knowledge of business and management principles involved in strategic planning, resource allocation, human resources modeling, leadership technique, production methods, and coordination of people and resources.
- *English language*—Knowledge of the structure and content of the English language including the meaning and spelling of words, rules of composition, and grammar.
- *Education and training*—Knowledge of principles and methods for curriculum and training design, teaching and instruction for individuals and groups, and the measurement of training effects.
  - Service orientation—actively looking for ways to help people.
  - Complex problem solving—identifying complex problems and reviewing related information to develop and evaluate options and implement solutions.
  - Coordination—adjusting actions in relation to others' actions.
  - Critical thinking—using logic and reasoning to identify the strengths and weaknesses of alternative solutions, conclusions, or approaches to problems.
  - Speaking—talking to others to convey information effectively.
  - Deductive reasoning—the ability to apply general rules to specific problems to produce answers that make sense.
  - Oral comprehension—the ability to listen to and understand information and ideas presented through spoken words and sentences.
  - Oral expression—the ability to communicate information and ideas in speaking so others will understand.
  - Problem sensitivity—the ability to tell when something is wrong or is likely to go wrong. It does not involve solving the problem, only recognizing there is a problem.
  - Speech clarity—the ability to speak clearly so others can understand you.

Having the ability and communicating skills with supervisors, peers, or subordinates will provide the information by using the following means, telephone, written form, e-mail, or in person.

Communicating with supervisors, peers, or subordinates—providing information to supervisors, coworkers, and subordinates by telephone, in written form, e-mail, or in person.

Getting information—observing, receiving, and otherwise obtaining information from all relevant sources.

Making decisions and solving problems—analyzing information and evaluating results to choose the best solution and solve problems.

Communicating with persons outside the organization—communicating with people outside the organization, representing the organization to customers, the public, government, and other external sources. This information can be exchanged in person, in writing, or by telephone or e-mail.

Establishing and maintaining interpersonal relationships—developing constructive and cooperative working relationships with others, and maintaining them over time.

- Coordinate special events or programs.
- Develop emergency response plans or procedures.
- Establish interpersonal business relationships to facilitate work activities.
- Maintain knowledge of current developments in area of expertise.
- Evaluate program effectiveness.

CONSIDERABLE PREPARATION NEEDED

Education: Is really important and seeking a degree in the emergency management field is an added plus.

The required experience and work related skills are necessary to gain the required knowledge for this occupation.

Job training: Emergency management on the job training and experiences in these occupations usually need several years of work-related experience, on-the-job training, and/or vocational training.

Job zone: Many of these occupations involve coordinating, supervising, managing, or training others.

Training certifications:

1. Emergency Management Institute (EMI) I National Preparedness Directorate National Training and Education Division:
   a. FEMA-based certification
      - The FEMA certification programs are conducted in cooperation with the Emergency Management Institute (EMI). The first certification is the Professional Development Series (PDS). It is awarded upon completion of seven FEMA courses. These courses may be taken as independent study or in the classroom.

   The required courses in the PDS program are the following:
      - Principles of Emergency Management, IS230
      - Emergency Planning, IS235
      - Effective Communication, IS242
      - Decision-Making and Problem-Solving, IS241
      - Leadership and Influence, IS240
      - Developing and Managing Volunteers, IS244
      - Exercise Design, IS139

   Once the PDS program is completed, you can then consider going on to the Advance Professional series (APS) certification, but you have to submit a request to enroll in the APS program through your State Emergency Manager Training Officer. If approved, your request will then be forwarded to your FEMA regional office. The next step is final approval by the Superintendent of the Emergency Management Institute (EMI).

The required APS courses are as follows:

- EOC Management and Operations, G275
- Incident Command System/Emergency Operations Center Interface, G191
- Rapid Assessment Workshop, G250.7
- Recovery from Disaster: The Local Government Role, G270.4
- Mitigation Planning Workshop for Local Governments, G318
- Then select five of these elective courses:
- Donations Management Workshop, G288
- Multi-Hazard Emergency Planning for Schools, G362
- Emergency Planning and Special Needs Populations, G197
- Resource Management, G276
- Debris Management, G202
- Mass Fatalities, G386
- Exercise Program Manager, G137
- Flood Fight Operations, G361
- Emergency Management Operations Course for Local Governments, G110
- Homeland Security Planning for Local Governments, 408
- Community Mass Care Management, G108
- Evacuation and Re-Entry Planning, G358
- Basic Public Information Officer, G290
- Hazardous Weather and Flood Preparedness, G271
- Warning Coordination, G272
- Advanced Incident Command System, G196

CERTIFIED EMERGENCY MANAGER

The Certified Emergency Manager (CEM) is considered by many to be the top certification in emergency management (IAEM, 2017). Created as a joint venture by FEMA and the International Association of Emergency Managers (IAEM), The CEM was developed as a standard to recognize professional competency in emergency management across the nation. Since the creation of the CEM, it has spread worldwide with the 1,000th CEM being certified within the last year.

The requirements for the Certified Emergency Manager Program are as follows:

- Three years emergency management experience
- A four-year baccalaureate degree; those without the education requirement should inquire about the Associate Emergency Manager (AEM) credential.
- 100 contact hours in emergency management training and 100 hours in general management training
- Six separate contributions to the profession, such as being published, speaking, professional membership, or other activities beyond your day-to-day activities

- A comprehensive emergency management essay to display your knowledge in all phases of emergency management
- Completion of a 100 question multiple-choice examination involving all topics in emergency management
- The requirements for the AEM program are the same as the CEM programs, except applicants are not required to hold a baccalaureate degree.

Maintaining Certification

The PDS and APS do not require a recertification once initially completed. The CEM/AEM credential requires recertification every five years. That recertification is required to remain current in the emergency management field. The recertification requirements include 100 hours of training, with at least 75 hours in emergency management and six contributions to the field of emergency management.

Benefits of Certifications

You may be asking yourself what the benefits of being certified are. One of the primary reasons for becoming certified is to be NIMS compliant. By being certified, you show that you are knowledgeable in the four core components of emergency management: preparedness, response, recovery, and mitigation. The certification is recognized as a standard that is recognized across the country. Another reason is to be professionally recognized by other professional emergency managers. The next reason may appear to be a little selfish, but it is valid nevertheless. By being certified, you are more marketable in looking for another position, whether it is for a promotion or another position outside of the law enforcement profession. Many positions for managers in emergency management or homeland security are seeking a CEM as a prerequisite to employment. With that in mind, when will you start your certification process?

Some really important traits for an emergency manager or those working in the emergency management field are the following:

- *Dependability*—Job requires being reliable, responsible, and dependable, and fulfilling obligations.
- *Integrity*—Job requires being honest and ethical.
- *Stress tolerance*—Job requires accepting criticism and dealing calmly and effectively with high-stress situations.
- *Initiative*—Job requires a willingness to take on responsibilities and challenges.
- *Leadership*—Job requires a willingness to lead, take charge, and offer opinions and direction.
- *Independence*—Occupations that satisfy this work value allow employees to work on their own and make decisions. Corresponding needs are Creativity, Responsibility, and Autonomy.

- *Relationships*—Occupations that satisfy this work value allow employees to provide service to others and work with co-workers in a friendly, noncompetitive environment. Corresponding needs are Coworkers, Moral Values, and Social Service.
- *Recognition*—Occupations that satisfy this work value offer advancement and potential for leadership and are often considered prestigious. Corresponding needs are Advancement, Authority, Recognition and Social Status.

In closing, the overall responsibility of an emergency manger is to maintain public trust, and confidence is central to the effectiveness of the emergency management profession. The emergency manager must reflect the spirit and proper conduct dictated by the conscience of society and commitment to the well-being of all and abide by the core values of respect, commitment, professionalism, ethics, and integrity.

A really good tool to follow as an emergency manager is the Code of ethics at the IAEM. Their website is http://www.iaem.com/page.cfm?p=certification/cem-code-of-ethics.

16 School Emergency Response Plan Template

EMERGENCY RESPONSE PLAN (ERP) TEMPLATE

The Emergency Response Plan (ERP) Template supplied in this chapter provides guidance that can be used by your district and schools in developing or revising each site's own comprehensive Emergency Response Plan (ERP). The template may be used in its entirety or in part. Districts and schools should work together to adapt the template to their own unique needs.

The process of developing, implementing, and maintaining a viable all hazards response capability is derived from the Federal Emergency Management Agency (FEMA). FEMA developed the Comprehensive Preparedness Guide (CPG) 101, Version 2, and November 2010 (https://www.fema.gov/media-library-data/20130726-1828-25045-0014/cpg_101_comprehensive_preparedness_guide_developing_and_maintaining_emergency_operations_plans_2010.pdf), which includes key elements of a viable ERP that have been incorporated in this template. The minimum requirements for school ERPs, per ARS 15-341 Part A. 32, may be found in the attached Minimum Requirements document.

In the current format, this template is public information; however, districts and schools should be aware of the need to protect specific emergency planning information and should redact information from the final ERP as necessary for the safety of the school.

EMERGENCY RESPONSE PLAN ORGANIZATION

An Emergency Response Plan (ERP) contains policies and procedures to follow before, during and after an emergency. The ERP integrates emergency preparedness activities into one document. It is the focal point for school planning and preparedness procedures.

Upon the recommendation of the Arizona Department of Emergency and Military Affairs, Division of Emergency Management and local emergency management, the AGENCY/DEPARTMENT–FOCUSED ERP FORMAT, found on page 3–8 of the Federal Emergency Management Agency (FEMA), Comprehensive Preparedness Guide 101 (CPG 101), Version II, November 2010, was selected for this ADE ERP template. This format supports key elements of a viable school ERP. Although, we have chosen this particular format to use as our template, CPG 101 offers a variety of formats that you may select from for your school emergency response plan.

Additional tools to aid in developing and implementing an ERP are found at http://www.azed.gov/prevention-programs/resources/#11schoolresourse.

AGENCY/DEPARTMENT–FOCUSED ERP FORMAT BASIC PLAN

SECTION I:

a) INTRODUCTION MATERIALS
   i. Promulgation Document/Signatures
   ii. Approval and Implementation
   iii. Record of Changes
   iv. Record of Distribution
   v. Table of Contents

b) PURPOSE, SCOPE SITUATION OVERVIEW AND ASSUMPTIONS
   i. Purpose
   ii. Scope
   iii. Situation Overview
      a. Hazard Analysis Summary
      b. Capability Assessment
      c. Mitigation Overview
   iv. Planning Assumptions
   v. Concept of Operations
   vi. Organization and Assignment of Responsibilities
   vii. Direction, Control, and Coordination
   viii. Information Collection, Analysis and Dissemination
   ix. Communication
   x. Administration, Finance and Logistics
      Plan Development and Maintenance
   xi. Authorities and References

SECTION II: LEAD AGENCIES
   i. FIRE
   ii. LAW ENFORCEMENT
   iii. EMERGENCY MEDICAL
   iv. EMERGENCY MANAGEMENT
   v. HOSPITAL
   vi. PUBLIC HEALTH
   vii. OTHERS AS NEEDED

SECTION III: SUPPORT AGENCIES

Identify those agencies that have a support role during an emergency and describe or address the strategies they are responsible for implementing.

SECTION IV: HAZARD-SPECIFIC PROCEDURES

For any response or support agency, describe or address its hazard-specific strategies.

EXAMPLE 1 LERP FOR SCHOOLS

Table of Contents

I. INTRODUCTION MATERIAL
   A. PROMULGATION STATEMENT
   B. APPROVAL AND IMPLEMENTATION

I. INTRODUCTION MATERIAL

A. Promulgation Statement

Promulgation Document

Promulgation is the process that officially announces/declares a plan (or law). The promulgation document gives the plan official status. It gives both the authority and the responsibility to organizations to perform their tasks.

It should also mention the responsibilities of tasked organizations with regard to preparing and maintaining their own procedures/guidelines and commit those organizations to carrying out the training, exercises, and plan maintenance needed to support the plan. In addition, the promulgation document allows senior officials to affirm their support for emergency management.

The Emergency Response Plan (ERP) and supporting materials, is a guide to how the School conducts all-hazards response. To clarify, all-hazards events (please refer to ERP Section IV: Hazard Specific Procedures) are typically associated with the following categories:

- Natural Hazard—These events are emergencies caused by forces extraneous to man in elements of the natural environment (e.g., earthquake, flood, hazardous weather, public health emergency).
- Technological Hazard—These incidents involve materials created by man and that pose a unique hazard to the general public and environment. The jurisdiction needs to consider incidents that are caused by accident (e.g., mechanical failure, human mistake), result from an emergency caused by another hazard, or are caused intentionally (e.g., infrastructure/utility disruption, radiological, or hazardous material release).
- Human-caused—These are disasters created by man, either intentionally or by accident (e.g., criminal or violent behavior, intruder, civil unrest, active shooter, terrorism).

More guidance on all-hazards planning may be found in Section IV: Hazard Specific Procedures.

The ERP is written in support of emergency management and is built upon the National Response Framework as scalable, flexible and adaptable coordinating structures to align key roles and responsibility. This plan and contents within shall apply to all (add School/District name) administration, staff, and students, and others participating in protection, prevention, mitigation, preparedness, response and recovery efforts. Furthermore, tasked organizations supporting ERP procedures shall maintain their own procedures and actively participate in the training, exercise, and maintenance needed to support this plan.

This plan is intended to capture specific authorities and best practices for managing incidents from the serious but purely local, to large-scale community or regional events, or catastrophic in nature.

Most emergencies follow some recognizable build-up period during which actions can be taken to achieve an appropriate state of maximum readiness. General actions are detailed in the appropriate sections of this document; however, it is acknowledged that disasters are unique occurrences, which require specific resources dependent upon the type, nature, and extent of the emergency. In this regard, this document is not all-inclusive, nor does it limit or restrict reasonable or prudent actions.

This ERP was prepared by (add School/District name) staff and approved by senior administration, whereas enabling activities contained within this document to be performed within the school capability.

Furthermore, this ERP has been distributed internally within the (add School/District name) and with external agencies that may be affected by its implementation.

B. Approval and Implementation

Approval and Implementation Page

The approval and implementation page introduces the plan, outlines its applicability, and indicates that it supersedes all previous plans. It should include a delegation of authority for specific modifications that can be made to the plan and by whom they can be made without the senior official's signature. It should also include a date and should be signed by the senior official(s) (e.g., governor, tribal leader[s], mayor, county judge, commissioner[s]).

This plan shall apply to all (add School/District name) students, faculty, and staff and others participating in protection, prevention, mitigation, preparedness, response and recovery efforts. Furthermore, the ERP may be applied to any school-sponsored events whether on or off property and all public or private, school-sanctioned activities.

The School/District (add Principal, Superintendent, etc.), or designee shall be responsible for plan oversight and coordination with applicable stakeholders, such as, law enforcement, county health department, fire services, emergency management, etc. This emergency response plan is based on the "all-hazards" concept and plans for natural and man-made disasters and incidents. The plan is flexible in that part of the plan or the entire plan may be activated based on the specific emergency and decision by school leadership.

This ERP and its supporting contents are hereby approved, supersedes all previous editions formerly referred to as the (insert previous name of document), and is effective immediately upon the signing of signature authority noted below.

Approval Signature: ______________________________,
Principal
School Name: ______________________________
Date: ______________________________

DISTRICT LEVEL SIGNATURE PAGE

Approval Signature: ______________________________,
Superintendent
District Name: ______________________________
Date: ______________________________

The following signatures appear in recognition of the coordination, collaboration, and document review with local partners.

Signature: ____________________ Date: ____________________
(Emergency Management Agency)
Signature: ____________________ Date: ____________________
(Public Health Department)
Signature: ____________________ Date: ____________________
(Law Enforcement Agency)
Signature: ____________________ Date: ____________________
(Fire Services Agency)

II. PURPOSE, SCOPE, SITUATION, AND ASSUMPTIONS

A. Purpose

Purpose

The purpose sets the foundation for the rest of the EOP. The basic plan's purpose is a general statement of what the EOP is meant to do. The statement should be supported by a brief synopsis of the basic plan and annexes.

The purpose of the (add School/District name) Emergency Response Plan is to outline the school's approach to emergency operations and to enable local, State and federal coordination in addition to public/private support. It provides general guidance for emergency management activities. The ERP and its contents describe the school emergency response organization and assigns responsibilities for various emergency tasks.

Specific support materials found in the ERP Sections beyond the Basic Plan, or in attachments, may describe details of who does what, when and how, or provides additional information in support of potential emergency response actions.

Additionally, the ERP describes capabilities and resources, as well as establishes responsibilities and operational processes, to help protect the school from natural, technological, and man-made hazards; with the primary objectives to save lives and protect public health, property and the environment, and, reduce adverse psychological consequences and disruptions.

Although emergencies/disasters and catastrophic incidents typically result in impacts far beyond the immediate or initial incident area, the ERP provides a framework to enable the management of cascading impacts and multiple incidents as well as the prevention of and preparation for subsequent events.

The (add School/District name) Emergency Response Plan supports the school and the district general operating procedures. It is the responsibility of those referenced in this plan to integrate their departmental policy, procedures, and emergency management activities such as task performance and organization, while also participating in training, exercises and plan integration and maintenance needed to support a collective process.

Additionally, the Plan:

- Empowers employees in an incident to act quickly and knowledgably;
- Informs students, faculty, and staff, and trains key stakeholders on their roles and responsibilities before, during, and after an incident;
- Provides other members of the community with assurances that the School/District has established policy and procedures to respond to incidents/hazards in an effective way;
- Establishes intra-agency and multi-jurisdictional mechanisms for involvement in, and coordination of, incident response and recovery operations;
- Provides guidance for emergency operations and the utilization of all available (add School/District name) and government resources for the protection of lives, property, and the continuance of school operations in an emergency.

B. Scope

Scope

The ERP/EOP should also explicitly state the scope of emergency and disaster response and the entities (e.g., departments, agencies, private sector, citizens) and geographic areas to which the plan applies.

It is the school's (add staff position, or designee) that is the responsible authority to direct the response involving an incident occurring on property, or at any (add School/District name) event sponsored or sanctioned.

This plan and all contents contained herein shall apply to all (add School/District name) students, faculty, staff and visitors, and others participating in protection, prevention, mitigation, preparedness, response and recovery efforts. An incident or event influencing (add School/District name) may also affect the surrounding community; therefore, this plan shall support community emergency operations and any supporting Memorandums of Understanding (MOU) and/or Memorandums of Agreement (MOA).

The scope of the ERP applies to incidents and/or events of all sizes, including those that exceed the school response services and capabilities that may result in activation of the local emergency operations center. MOUs/MOAs in coordination with additional support requested from local, State and Federal entities may be activated.

The School ERP outlines the expectations of faculty and staff; roles and responsibilities; direction and control systems; internal and external communications; training and sustainability of the ERP; authority and references as defined by local, State, and Federal government mandates; common and specialized procedures; and specific hazard vulnerabilities and response/recovery for (add School/District name).

C. Situation Overview

Situation Overview

The situation section characterizes the "planning environment," making it clear why an EOP is necessary. The level of detail is a matter of judgment; some information may be limited to a few specific annexes and presented there.

At a minimum, the situation section should summarize hazards faced by the jurisdiction and discuss how the jurisdiction expects to receive (or provide) assistance within its regional response structures. The situation section covers a general discussion of:

- Relative probability and impact of the hazards
- Geographic areas likely to be affected by particular hazards
- Vulnerable critical facilities (e.g., nursing homes, schools, hospitals, infrastructure)
- Population distribution and locations, including any concentrated populations of individuals with disabilities, others with access and functional

needs, or individuals with limited English proficiency, as well as unaccompanied minors and children in daycare and school settings
- Dependencies on other jurisdictions for critical resources
- The process used by the jurisdiction to determine its capabilities and limits in order to prepare for and respond to the defined hazards
- Actions taken in advance to minimize an incident's impacts, including short- and long-term strategies.

(Add School/District name) is a stakeholder in the local and state disaster risks. The school/district is exposed to hazards, which have the potential for disrupting the school community and causing widespread damage and casualties. Such hazard exposure may include geographic location, population concentration to include demographics, buildings, rail, air and highway traffic.

Other risk factors may include: floods, tornadoes, terrorist activities, fires, explosions, transportation accidents, pandemic and other infectious diseases, shooting, school collapse, tornado, bomb threats, hostage situation, utility outage, neighborhood disturbance, food poisoning, armed intruder, etc.

The information gathered during the jurisdictional assessment of individuals with disabilities and others with access and functional needs requires a detailed analysis. Emergency planners need to review the assessment findings and analyze the quantity and types of resources (including personnel) needed during different types of incidents.

The (add School/District name) participated in the hazard assessment process, utilizing local resources, such as emergency management, law enforcement, fire services, county health department and private entities where applicable, to determine the threat and risk for the school and surrounding community.

For information on (add School/District name) demographics, building information, threat and risk, please see (add School/District web site) or (add attachment location).

To enhance capability and mitigation efforts, the (add School/District name) is a signatory on (may be local emergency management plan, MOU, etc.), which may be found at the following location(s) (add locations). Capabilities to be able to respond to the most likely hazards were assessed by the school and the surrounding emergency management community. Collectively, measures are in place to address these hazards.

- The school works with the law enforcement, county health department, fire services and emergency management staff of the following city (ies) (add names) and/or county (ies) of (add names) on a regular basis. A cooperative working relationship and team approach between the school and municipal governments for emergency response is seen as a major responsibility for which the school supports. Prevention, protection, response, and recovery capability consideration, along with the adequate training, equipment, and personnel needs are coordinated.

- The school is dependent upon the local municipalities for fire suppression and rescue services, major hazardous material response capabilities, EMS, bomb squad response, public health guidance, law enforcement tactical responses, support from emergency management, and assistance in emergency operations staffing. If written agreements exist, add location of these agreements.
- The school maintains staffing resources, or has written agreements in place, that may provide the following services: (add items such as CPR, first aid, counseling).

The (add School/District name) has assigned the following staff, (list position) to address emergency preparedness. In addition to supporting (add local, private, and regional stakeholders) the school has adopted and supports FEMA emergency management practices, protection, prevention, mitigation, preparedness, response and recovery in their planning process, and is committed to developing and exercising plans in addition to training and exercising with local populations.

D. Planning Assumptions

Planning Assumptions

These identify what the planning team assumes to be facts for planning purposes in order to make it possible to execute the EOP. During operations, the assumptions indicate areas where adjustments to the plan have to be made as the facts of the incident become known. These also provide the opportunity to communicate the intent of senior officials regarding emergency operations priorities.

The (add School/District name) ERP is based on the following planning assumptions and considerations as presented in this section.

- Any employee of (add School/District name) may be tasked by this ERP.
- School officials and representatives recognize their responsibilities for the safety and well-being of students, staff, and visitors; and assume their responsibilities in the implementation of this ERP.
- External resources may be requested to assist the school.
- In most cases, law enforcement or fire service personnel will assume Incident Command, or establish a unified command, depending on the type of emergency.
- Incident management activities will be initiated and conducted using the Incident Command System, contained in the National Incident Management System (NIMS).
- Outside assistance will likely be available in most emergencies affecting the school. Although these procedures define procedures for coordinating such assistance, it is essential for the school to be prepared to carry out disaster response and short-term actions on an independent basis, or in the event, the incident is community or region wide.

- It is possible for a major disaster to occur any time and any place in or near the school. In some cases, timely dissemination of warnings and increased readiness measures may be possible. However, many disasters can, and may occur with little or no warning.
- Proper implementation and understanding of these procedures through training and exercising will reduce or prevent disaster-related losses.
- Emergencies on the School grounds may involve multiple responding departments and agencies including, but not limited to, local law enforcement, county health department, fire services, emergency management, environmental health and safety, emergency medical services, and appropriate school, city, county, state and federal agencies.
- Other schools operating on the (add School/District name) property shall coordinate their emergency actions with this ERP. (Add other items as applicable)

III. CONCEPT OF OPERATIONS

Concept of Operations (CONOPS)

The audience for the basic plan needs to be able to visualize the sequence and scope of the planned emergency response. The CONOPS section is a written or graphic statement that explains in broad terms the decision maker's or leader's intent with regard to an operation.

The CONOPS should describe how the response organization accomplishes a mission or set of objectives in order to reach a desired end-state. Ideally, it offers clear methodology to realize the goals and objectives to execute the plan. This may include a brief discussion of the activation levels identified by the jurisdiction for its operations center. The CONOPS should briefly address direction and control, alert and warning, and continuity matters that may be dealt with more fully in annexes.

A. General

It is the responsibility of the school to protect life and property from the effects of emergency situations within its own jurisdiction. (Add School/District name) has the primary responsibility for initial emergency management activities. Concept of Operations information located in this section is designed to give an overall picture of incident management.

It will primarily clarify the purpose, and explain the school's overall approach to an emergency (i.e., what should happen, when, and at whose direction).

Top priorities for incident management are to:

- Protect and save lives, and protect health and safety of students, faculty, staff, visitors, responders, and recovery workers.
- Protect property and mitigate damages and impacts to individuals, the community, and the environment.

To assist in delineating levels of response, the following terms have been provided, but it is of the utmost importance that terminology is used that is acceptable and applicable to your individual school, district and response community.

An emergency, as used in this plan, is intended to describe a range of situations, from an incident to a major disaster. It includes the following:

INCIDENT—An incident is a situation that is limited in scope and potential effects.

EMERGENCY—An emergency is a situation that is larger in scope and more severe in terms of actual or potential effects than an incident.

DISASTER—A disaster involves the occurrence or threat of significant casualties and/or widespread property damage that is beyond the capability of the local government to handle with available local resources.

This ERP is based on the "all-hazards" concept and is flexible in that part of the plan or the entire plan may be activated based on the specific emergency and decision by school leadership.

The school concept of operations is that the emergency functions of various departments and agencies involved in emergency management will generally parallel normal day-to-day functions or operations.

Day-to-day functions that do not contribute directly to the emergency may be suspended for the duration of any emergency. The efforts that would normally be required for those functions will be redirected to the accomplishment of emergency tasks by the school.

The (add School/District name) ERP addresses preparedness activities embedded within the key areas of emergency planning (CPG 101 v.2, pg. 1–8).

B. Six Key Areas of Emergency Planning and Incident Management

In the event of an incident, the (add School/District name) will utilize these definitions (CPG 101, pgs. 1–8/1–10) that are predicated on an all-hazard approach. There is acknowledgement that most responsibilities and functions performed during an emergency are not hazard specific. Likewise, these procedures account for activities pre-incident, incident, and post-incident; consequently, key areas are noted as the following:

1. Prevention—Consists of actions that reduce risk from human-caused incidents. Prevention planning can also help mitigate secondary or opportunistic incidents that may occur after the primary incident.
2. Protection—Reduces or eliminates a threat to people, property and the environment. Primarily focused on adversarial incidents, the protection of critical infrastructure and key resources (CIKR) is vital to local jurisdictions, national security, public health and safety and economic vitality.
3. Mitigation—Mitigation activities are those which eliminate or reduce the probability of a disaster occurring. Also included are those long-term activities which lessen the undesirable effects of unavoidable hazards.

4. Preparedness—Preparedness activities serve to develop the response capabilities needed in the event an emergency should arise. Planning and training are among the activities conducted under this phase.
5. Response—Response is the actual provision of emergency services during a crisis. These activities help to reduce casualties and damage, and speed recovery. Response activities include evacuation, rescue, and other similar operations.
6. Recovery—Recovery is both a short-term and long-term process. Short-term operations seek to restore vital services to the School and provide for the basic needs of students, faculty, and staff. Long-term recovery focuses on restoring the School to its normal pre-disaster, or an improved, state of affairs. The recovery period is also an opportune time to institute future mitigation measures, particularly those related to the recent emergency.

C. National Incident Management System (NIMS)

NIMS provides a consistent framework for incident management at all jurisdictional levels regardless of the cause, size, or complexity of the incident. NIMS is not an operational incident management or resource allocation plan. NIMS represents a core set of doctrine, concepts, principles, terminology, and organizational processes that enables effective, efficient, and collaborative incident management.

NIMS ensures that those involved in incident response and recovery, understand what their roles are and have the tools they need to be effective. Additionally NIMS:

- Defines incident response organizational concepts and structures.
- Consists of procedures for managing personnel, facilities, equipment, and communications.
- Is used throughout the life cycle of an incident (e.g., from threat to parent/student reunification).

NIMS components include:

- Preparedness—Effective emergency management and incident response activities begin with a host of preparedness activities conducted on an ongoing basis, in advance of any potential incident. Preparedness involves an integrated combination of planning, procedures and protocols, training and exercises, personnel qualifications and certification, and equipment certification.
- Communications and Information Management—Emergency management and incident response activities rely upon communications and information systems that provide a common operating picture to all command and coordination sites. NIMS describes the requirements necessary for a standardized framework for communications and emphasizes the need for a common operating picture. NIMS is based upon the concepts of interoperability, reliability, scalability, portability, and the resiliency and redundancy of communications and information systems.

- Resource Management—Resources (such as personnel, equipment, and/or supplies) are needed to support critical incident objectives. The flow of resources must be fluid and adaptable to the requirements of the incident. NIMS define standardized mechanisms and establishes the resource management process to: identify requirements, order and acquire, mobilize, track and report, recover and demobilize, reimburse, and inventory resources.
- Command and Management—The Command and Management component within NIMS is designed to enable effective and efficient incident management and coordination by providing flexible, standardized incident management structures. The structures are based on three key organizational constructs: the Incident Command System, Multiagency Coordination Systems, and Public Information.
- Ongoing Management and Maintenance—Within the auspices of Ongoing Management and Maintenance, there are two components: the National Integration Center (NIC) and Supporting Technologies.

The (add School/District name) has adopted NIMS and the use of the Incident Command System (ICS) in accordance with the Homeland Security Presidential Directive (HSPD) 5. Additionally, the U.S. Department of Education has provided guidance as to "key personnel," such as executive leaders, general personnel, Command Staff and Incident Managers, to complete specific courses in order to meet organizational NIMS compliance.

Appropriate school staff will participate in NIMS preparedness programs, community collaboration and planning efforts, and participate in training and exercising of the ERP's procedures.

In a major emergency or disaster, the school may be damaged or need to be evacuated, people may be injured, and/or other incident management activities may need to be initiated. These activities will be organized and coordinated to ensure efficient incident management. The Incident Command System (ICS), a component of NIMS, will be used to manage all incidents and major school/district planned events. The school ICS approach will be used in all phases of incident management, including pre-incident activities, incident and post incident.

D. Emergency Operations Plan Activation Authority and Communication

The (add at least two to three staff positions) is/are typically the responsible authority for directing emergency measures for the school or district, and is provided the authority to activate this ERP. (Add staff position order of succession here)

The (add school staff position(s)) will obtain information on the nature of the incident and assessment of the situation and will make the determination of next steps and assume the role of, or delegate the role of incident commander if the situation warrants.

An incident warranting multi-jurisdictional resources, emergency response activities will employ the Incident/Unified Command System (ICS) structure.

Timely warnings of emergency conditions are essential to preserve the safety and security of the school or district community and critical to an effective response and recovery. Direction on communications may be found (add location). Additional information may be obtained at the District level (add location).

IV. ORGANIZATION AND ASSIGNMENT OF RESPONSIBILITIES

Organization and Assignment of Responsibilities

The basic plan establishes the operational organization that will be relied on to respond to an emergency. It includes a list of the kinds of tasks to be performed, by position and organization, without all of the procedural details included in functional annexes. When two or more organizations perform the same kind of task, one should be given primary responsibility, with the other(s) providing a supporting role. For the sake of clarity, a matrix of organizations and areas of responsibility (including functions) should be included to summarize the primary and supporting roles. Shared general responsibilities, such as developing SOPs/SOGs, should not be neglected, and the matrix might include organizations not under jurisdictional control, if they have defined responsibilities for responding to emergencies that might occur in the jurisdiction. Organization charts, especially those depicting how a jurisdiction is implementing the ICS or Multiagency Coordination System structure, are helpful.

This section should also outline agency and departmental roles related to prevention and protection activities. In addition, this section is where a jurisdiction discusses the option that it uses for organizing emergency management—ESF, agency and department, functional areas of ICS/NIMS, or a hybrid. The selected management structure determines what types of annexes are included in the EOP and should be carried through to any hazard annexes.

A. Organization

Emergency Management for (add School/District name) is comprised of the following:

(The following are examples and should be modify as applicable to your school or district.)

Policy Group (District)

The Policy Group is comprised of the following: (add position names)

- Superintendent
- School Board President
- Other

In complex incidents, the Policy Group will be convened at (add location and alternative location). The role of the Policy Group is to:

- Support the on-scene Incident Commander with strategic guidance, information analysis, and needed resources
- Provide policy and strategic guidance
- Help to ensure that adequate resources are available
- Identify and resolve issues common to all organizations
- Keep elected officials and other executives informed of the situation and decisions
- Provide factual information, both internally and externally through the Joint Information Center (JIC)

School Site Safety and Security Staff

The (add position name) is responsible for (list what this position is responsible for on a daily bases. It may include duties such as emergency management planning and operations for the school.) Upon the activation of the ERP for a school incident, the (add position name) assumes the role as (list the role this position will assume), unless delegated.

District or School Departments

District and school departments will support emergency response operations through performance of their normal roles and responsibilities. If called upon, school departments will activate personnel and implement appropriate response actions identified in the plan, or as directed by the Incident Commander or school administration.

Activation of the ICS for a school event may be modified as needed to best serve the nature of the incident. When the ICS is activated, staff will direct the efforts of their departments according to their respective procedures for emergency operations.

Department	Departmental Duties and Responsibilities
1. Transportation	Upon learning of an incident, the Transportation Coordinator/Director will report directly to the Incident Commander for assignment unless otherwise notified.

B. School Position Roles and Expected Actions

Principal/Building Administrator

The principal may serve as the Incident Commander or delegate that authority to a qualified individual. At all times, the principal still retains the responsibility for the overall safety of students and staff. However, delegating the authority to manage the incident allows the principal to focus on policy-level activities and interfacing with other agencies and parents. The principal shall coordinate between the policy group and the Incident Commander.

Teachers

Teachers shall be responsible for the supervision of students and shall remain with students until directed otherwise.

Responsibilities include:

- Supervise students under their charge.
- Take steps to ensure the safety of students, staff, and other individuals in the implementation of incident management protocols.
- Direct students in their charge to inside or outside assembly areas, in accordance with signals, warning, written notification, or intercom orders according to established incident management procedures.
- Give appropriate action command during an incident.
- Take attendance when class relocates to an outside or inside assembly area or evacuates to another location.
- Report missing students to the Incident Commander or designee.
- Execute assignments as directed by the Incident Commander or ICS supervisor.
- Obtain first aid services for injured students from the school nurse or person trained in first aid. Arrange for first aid for those unable to be moved.
- Render first aid if necessary. School staff should be trained and certified in first aid and CPR.

Instructional Assistants

Responsibilities include assisting teachers as directed.

Counselors, Social Workers, and Psychologists

Counselors, social workers, and psychologists provide assistance with the overall direction of the incident management procedures at the site.

Responsibilities may include:

- Take steps to ensure the safety of students, staff, and other individuals in the implementation of incident management protocols.
- Direct students in their charge according to established incident management protocols.
- Render first aid if necessary.
- Assist in the transfer of students, staff, and other individuals when their safety is threatened by a disaster.
- Execute assignments as directed by the Incident Commander or ICS supervisor.

School Nurses/Health Assistants

Responsibilities include:

- Administer first aid or emergency treatment as needed.
- Supervise administration of first aid by those trained to provide it.
- Organize first aid and medical supplies.

Custodians/Maintenance Personnel

Responsibilities include:

- Survey and report building damage to the Incident Commander or Operations Section Chief.
- Control main shutoff valves for gas, water, and electricity and ensure that no hazard results from broken or downed lines.
- Provide damage control as needed.
- Assist in the conservation, use, and disbursement of supplies and equipment.
- Keep Incident Commander or designee informed of condition of school.

School Secretary/Office Staff

Responsibilities include:

- Answer phones and assist in receiving and providing consistent information to callers.
- Provide for the safety of essential school records and documents.
- Execute assignments as directed by the Incident Commander or ICS supervisor.
- Provide assistance to the principal and Policy/Coordination Group.
- Monitor radio emergency broadcasts.
- Assist with health incidents as needed, acting as messengers, etc.

Food Service/Cafeteria Workers

Responsibilities include:

- Use, prepare, and serve food and water on a rationed basis whenever the feeding of students and staff becomes necessary during an incident.
- Execute assignments as directed by the Incident Commander or ICS supervisor.

Transportation/Bus Drivers

Responsibilities include:

- Supervise the care of students if disaster occurs while students are in the bus.
- Transfer students to new location when directed.
- Execute assignments as directed by the Incident Commander or ICS supervisor.
- Transport individuals in need of medical attention.

Other Staff (Itinerant Staff, Substitute Teachers)

Responsibilities include reporting to the Incident Commander or ICS supervisor if requested or activated.

Students

Responsibilities include:

- Cooperate during emergency drills and exercises, and during an emergency situation.
- Learn to be responsible for themselves and others if the emergency situation warrants.
- Understand the importance of not being a bystander by reporting situations of concern to appropriate staff.
- Develop a general awareness of natural, technological, and human-caused hazards and associated prevention, preparedness, and mitigation measures.

Parents/Guardians

Responsibilities include:

- Understanding their roles during a school emergency.
- Encourage and support school safety, violence prevention, and incident preparedness programs within the school.
- Participate in volunteer service projects for promoting school incident preparedness.
- Provide the school with requested information concerning the incident, early and late dismissals, and other related release information.
- Practice incident management preparedness in the home to reinforce school training and ensure family safety.

V. DIRECTION, CONTROL, AND COORDINATION

Direction, Control, and Coordination

This section describes the framework for all direction, control, and coordination activities. It identifies who has tactical and operational control of response assets. Additionally, Direction, Control, and Coordination explain how multijurisdictional coordination systems support the efforts of organizations to coordinate efforts across jurisdictions while allowing each jurisdiction to retain its own authorities. This section also provides information on how department and agency plans nest into the EOP (horizontal integration) and how higher-level plans are expected to layer on the EOP (vertical integration).

A. Incident Command

Incident Command System

The Incident Command System (ICS) organizational structure develops in a top-down, modular fashion that is based on the size and complexity of the incident, as well as the specifics of the hazard environment created by the incident. As incident complexity increases, the organization expands from the top down as functional responsibilities are delegated.

To provide for the effective direction, control, and coordination of an incident, either single site or multi-incidents, the school ERP will be activated including the implementation of the Incident Command System (ICS). When needed, separate functional elements will be established and subdivided to enhance internal organizational management and external coordination.

The Policy Group is responsible for providing the Incident Commander with strategic guidance, information analysis, and needed resources. The Executive/Senior Leadership (Principal, Superintendent, etc.) is responsible for the incident. Along with this responsibility, by virtue of their position, these individuals have the authority to make decisions, commit resources, obligate funds, and command the resources necessary to protect the students and facilities.

Having the responsibility does not mean that the Executive/Senior Leadership assumes a command role over the on-scene incident operation. Rather, the Executive/Senior Official:

- Provides policy guidance on priorities and objectives based on situational needs and the Emergency Operations Plan.
- Oversees resource coordination and support to the on-scene command from an Operations Center.

Incident Management

The school ICS may be organized into the following functional areas: provides strategic guidance and resource support.

Organization Expansion

Expanding the Organization

The School/District Incident Commander will be integrated into the Incident Command structure, or assume a role within a Unified Command structure.

If a school emergency is within the authorities of the first-responder community, i.e. emergency requiring law enforcement or fire services, etc., Command will transition, or form a Unified Command structure, upon the arrival of qualified first responders. A transfer of command briefing shall occur.

Incident Command Post

The Incident Command Post (ICP) is the location from which the Incident Commander oversees all incident operations. There is generally only one ICP for each incident, but it may change locations during the event. Every incident must have some form of an ICP.

The ICP will be positioned outside of the present and potential hazard zone, but located within safe proximity to the emergency site. The ICP is generally responsible for incident response management as follows:

- Serves as a temporary field location for tactical-level on-scene incident command and management
- Is the on-site headquarters for the Incident Commander, Command Staff and General Staff; and
- Serves as a field collection point for tactical intelligence and analysis
- The Incident Command Post (ICP) conducts all operations using the Incident/Unified Command System (ICS)
- The Incident Command Post provides the initial securing of the perimeter of the area, coordinates the actions of the operating units, and remains operational during the field actions (rescue, response, recovery, etc.) as required

Incident Commander

The Incident Commander (IC) is the primary person in charge at the incident and will establish incident objectives based on the following five primary objectives:

1. Life Safety
2. Protect Public Health
3. Incident Stabilization
4. Property and Environment Preservation
5. Reduce adverse psychological consequences and disruptions

Additionally, the IC will manage the incident scene, and he or she must keep the Executive/Senior Administration and the Policy Group informed and up to date on important matters pertaining to the incident.

School-related responsibilities and duties include:

- The first staff person on scene, (or insert position) will assume the role of Incident Commander until a more qualified individual can assume command.
- The Incident Commander is delegated the appropriate authority to direct tactical on-scene operations until a coordinated incident management framework can be established with local resources.
- Establish an Incident Command Post (ICP) and provide an assessment of the situation to the Executive/Senior Administration, which may also include the Policy Group or other officials, recommend incident response activities, identify incident management resources required, and direct the on-scene incident management activities from the ICP.
- Establish and manage the Command Post, establish the incident organization, and determine strategies to implement protocols and adapt as needed.
- Monitor incident safety conditions and develop measures for ensuring the safety of building occupants (including students, staff, volunteers, and responders).
- Coordinate media relations and information dissemination with the principal.
- Serve as the primary on-scene contact for outside agencies assigned to the incident, establish unified command as necessary, develop working knowledge of local/regional agencies, and assist in accessing services when the need arises.

- Document activities.
- Assume overall direction of all incident management procedures based on actions and procedures outlined in this EOP.
- Determine whether to implement incident management protocols (e.g., Evacuation, Shelter in Place, Lockdown, etc.), as described more fully in the (add location) in this document.
- Arrange for transfer of students, staff, and other individuals when safety is threatened by a disaster.
- Work with emergency services personnel. (Depending on the incident, community agencies such as law enforcement or fire department may have jurisdiction for investigations, rescue procedures, etc.)
- Keep the Executive/Senior Leadership and Policy Group informed of the situation.

Unified Command

Unified Command applies ICS to incidents involving multiple jurisdictions or agencies. It enables schools and agencies with different legal, geographic, and functional responsibilities to coordinate, plan, and interact effectively.

Additionally, the Incident Commanders within the Unified Command will make joint decisions and speak as one voice. Any differences are worked out within the Unified Command. Within Unified Command, law enforcement would not tell school personnel how to manage, i.e., parent-student reunification, etc., or tell the firefighters how to do their job.

- The Incident Commander, through the Incident/Unified Command System, coordinates the actions of responding community units to the scene
- Advise School Leadership and the District of needs that may include personnel recall from other departments/schools as required
- Isolate the incident site and maintain control of the inner and outer perimeters
- Establish tactical communications and designate a primary radio channel
- Facilitate tactical planning and contingency planning
- Brief first responder personnel
- Designate a staging area for supporting agencies
- Ensure documentation of decisions and activities
- Provide situational updates to the policy group
- Approve requests for additional resources or for the release of resources (demobilization)
- Approve additional alerts as needed
- Ensure staff prepares an incident After Action Report (AAR)
- Establish immediate priorities
- Coordinate any specific transportation issues (such as helicopter landing zones, EMS locations, morgue location, etc., as appropriate)
- Determines security boundaries
- Perform other duties as required by the situation

Public Information Officer

The Public Information Officer (PIO) is responsible for interfacing with the media or other appropriate agencies requiring information directly from the incident.

- Check in with Incident Commander and receive a situation brief
- Assume the duties of the School Public Information Officer (PIO)
- If necessary, establish and supervise a Joint Information Center (JIC) with PIO's from the other responder agencies
- Coordinate press releases among response organizations
- Designate a media center and facilitate scheduled press briefings
- Ensure all press releases and public information are reviewed and approved by the Incident Commander, or designee
- Monitor news media outlets reports of the incident
- Prepare periodic briefings to Executive Policy Group on public information activities and submit draft press releases for review and approval

Operations Section

The Operations Section directs all tactical operations of an incident including implementation of response/recovery activities according to established incident management procedures and protocols, care of students, first aid, crisis intervention, search and rescue, site security, damage assessment, evacuations, and the release of students to parents.

Specific responsibilities include:

- Analyze school staffing to develop a Parent-Student Reunification Plan, and implement an incident action plan.
- Monitor site utilities (i.e., electric, gas, water, heat/ventilation/air conditioning) and shut off only if danger exists or directed by Incident Commander, and assist in securing facility.
- Establish medical triage with staff trained in first aid and CPR, provide and oversee care given to injured persons, distribute supplies, and request additional supplies from the Logistics Section.
- Provide and access psychological first aid services for those in need, and access local/regional providers for ongoing crisis counseling for students, staff, and parents.
- Coordinate the rationed distribution of food and water, establish secondary toilet facilities in the event of water or plumbing failure, and request needed supplies from the Logistics Section.
- Document all activities.

Planning Section

Collects, evaluates, and disseminates information needed to measure the size, scope, and seriousness of an incident, and to plan appropriate incident management activities.

Duties may include:

- Assist Incident Commander in the collection and evaluation of information about an incident as it develops (including site map and area map of related events), assist with ongoing planning efforts, and maintain incident time log.
- Document all activities.

Logistics Section

Supports incident management operations by securing and providing needed personnel, equipment, facilities, resources, and services required for incident resolution; coordinating personnel; assembling and deploying volunteer teams; and facilitating communication among incident responders. This function may involve a major role in an extended incident.

Additional responsibilities include:

- Establish and oversee communications center and activities during an incident (two-way radio, battery-powered radio, written updates, etc.), and develop telephone tree for after-hours communication.
- Establish and maintain school and classroom preparedness kits, coordinate access to and distribution of supplies during an incident, and monitor inventory of supplies and equipment.
- Document all activities.

Finance/Administration Section

Oversees all financial activities including purchasing necessary materials, tracking incident costs, arranging contracts for services, timekeeping for emergency responders, submitting documentation for reimbursement, and recovering school records following an incident.

Additional duties may include:

- Assume responsibility for overall documentation and recordkeeping activities; when possible, photograph or videotape damage to property.
- Develop a system to monitor and track expenses and financial losses, and secure all records.

The Finance and Administration Section may not be established onsite at the incident. Rather, the school and school district management offices may assume responsibility for these functions.

An important component of the ERP is a set of interagency agreements with various city/county agencies to aid timely communication. These agreements help coordinate services between the agencies and (add District name). Various agencies and services include county governmental agencies such as mental health, law enforcement, county health department, and fire departments. The agreements specify the type of communication and services provided by one agency to another.

The agreements also make school personnel available beyond the school setting in an incident or traumatic event occurring in the community.

Source and Use of Resources

(Add School/District name) will use its own resources and equipment to respond to incidents until incident response personnel arrive. Parent volunteers and community members can be trained to assist if called upon and available after an incident occurs. The following organizations or agencies have agreed to be responsible for providing additional resources or assistance by means of written or contractual agreement: (add location of agreements)

Examples:

- First aid kit and sanitation supplies
- Counseling services
- Food/water supplies
- Security

B. Communication

Communications

This section describes the communication protocols and coordination procedures used between response organizations during emergencies and disasters. It discusses the framework for delivering communications support and how the jurisdiction's communications integrate into the regional or national disaster communications network. It does not describe communications hardware or specific procedures found in departmental SOPs/SOGs. Planners should identify and summarize separate interoperable communications plans. This section may be expanded as an annex and is usually supplemented by communications SOPs/SOGs and field guides.

Communication is a critical part of incident management. This section outlines (add School/District name) communications plan and supports its mission to provide clear, effective internal and external communication between the school, staff, students, parents, responders, and media. (Add location of communication plan, policy, and procedures.)

C. Notification and Warning

Timely warnings of emergency conditions are essential to preserve the safety and security of the school community and critical to an effective response and recovery.

- Upon learning of an emergency and assessing need for local agency support such as law enforcement or fire services, (add staff positions) will call 911.

- Notification to key (add School/District name) administrators, departments and personnel for emergency response will follow procedures outlined in (add your notification procedures here, or in a section such as an Appendix or Attachment to the ERP).

Notification of critical personnel will be in accordance with the following:

- District and/or school personnel shall relay threat information, warnings, to ensure the school community is notified.
- Staff shall respond according to their procedures for emergency operations, unless otherwise dictated by the event.
- Emergency notifications, warnings and alerts will typically be disseminated within the school or district using items such as, voice command, intercom system, email, the school radio or television system, or other modes.
- Law enforcement agencies and other emergency services may be required to disseminate emergency warnings to the public who cannot be reached by school primary warning systems.
- In any case, it is important for the communication hubs to be notified of the emergency to ensure all appropriate notifications are made. Any one or more communication strategies may become disabled.
- Telephones, cellular or landline, (add items applicable) are the primary means of communications for contacting key emergency responders or departments.
- The 800 or 900 MHz radios (if applicable) with common channels are one example of communication for emergency responder communications.
- All school radios (if applicable) have a set of common channels for interoperability among departments.

D. Emergency Public Information

In the event that a crisis occurs on school property, the Public Information Officer will be notified as soon as possible to report to the Incident Command location. In his/her role as chief spokesperson for the School, will ultimately be responsible for the communications efforts relative to the crisis.

For a detailed plan of emergency public information, (add location if applicable).

E. Non-Emergency External Communications

During an incident the school expects to receive a high volume of calls seeking information as to the welfare of students, staff and faculty from concerned parents, relatives, spouses, friends and loved ones. The surge in volume of calls to the school main numbers may quickly exceed the system capabilities.

It is essential that call centers are activated and staffed as soon as possible to handle anticipated volume of non-emergency calls related to the incident.

Call centers may also be a resource in helping to control rumors.

VI. INFORMATION COLLECTION, ANALYSIS, AND DISSEMINATION

Information Collection, Analysis, and Dissemination

This section describes the critical or essential information common to all operations identified during the planning process. It identifies the type of information needed, the source of the information, who uses the information, how the information is shared, the format for providing the information, and any specific times the information is needed. School/District prevention and protection assets must develop the Information Collection, Analysis, and Dissemination section in close cooperation with school departments and local support agencies such as, law enforcement, fire, emergency management, utilities, insurance agencies, risk management, transportation, etc. The contents of this section are best provided in a tabular format, and may be expanded as an annex if needed.

Essential information necessary for emergency operations identified in this ERP are recorded on Attachment 1. (If, Attachment 1 was not used, add actual process used) and shared with all appropriate school and district departments, and with external partners including law enforcement, fire, and emergency management.

More in-depth information in this topic area is found in the following materials:

- FEMA-428/BIPS-07/January 2012 Edition 2—this primer focuses on a single facility type with a very specific occupancy and vulnerability.
- FEMA P-424, Design Guide for Improving School Safety in Earthquakes, Floods, and High Winds (2010)—addresses the protection of schools from school shooting or terrorist threats.
- BIPS 06 (Formerly FEMA 426), Reference Manual to Mitigate Potential Terrorist Attacks against Buildings—deals with all building types and occupancies, and terrorism.

VII. ADMINISTRATION, FINANCE, AND LOGISTICS

Administration, Finance, and Logistics

This section covers general support requirements and the availability of services and support for all types of emergencies, as well as general policies for managing resources. Planners should address the following in this section of the plan:

- References to mutual aid agreements.
- Authorities for and policies on augmenting staff by reassigning public employees and soliciting volunteers, along with relevant liability provisions.
- General policies on keeping financial records, reporting, tracking resource needs, tracking the source and use of resources, acquiring ownership of resources, and compensating the owners of private property used by the jurisdiction.

If this section is expanded, it should be broken into individual functional annexes—one for each element.

Consider adding a section covering general support requirements and the availability of services and support for all types of emergencies, as well as general policies for managing resources and adding items such as:

- References to Mutual Aid Agreements: Written agreements between organizations, either public or private, for reciprocal aid and assistance in case of disasters too great may be dealt with unassisted.
- Authorities for, and policies on augmenting staff by reassigning public employees and soliciting volunteers, etc.
- General policies on keeping financial records, reporting, tracking resource needs, tracking the source and use of resources, acquiring ownership of resources, and compensating the owners of private property used by the school.

For the purposes of potential insurance, local, state, or federal assistance, or reimbursement, identify the process for which school or district event documentation is tracked. For example, Financial Management may issue a project number for the incident response effort, and may disseminate the project number for use by all school or district departments participating.

This project number would be utilized in conjunction with the applicable accounting code to document all response and recovery costs associated with any emergency or disaster requiring a substantial response effort.

VIII. PLAN DEVELOPMENT AND MAINTENANCE

Plan Development and Maintenance

This section discusses the overall approach to planning and the assignment of plan development and maintenance responsibilities. This section should:

- Describe the planning process, participants in that process, and how development and revision of different "levels" of the EOP (basic plan, annexes, and SOPs/SOGs) are coordinated during the preparedness phase.
- Assign responsibility for the overall planning and coordination to a specific position.
- Provide for a regular cycle of training, evaluating, reviewing, and updating of the EOP.

A. Plan Development and Maintenance

The ERP integrates with school and district policy and procedures and a number of stakeholder ERPs or guidelines. The school ERP utilizes existing program expertise and personnel to provide prevention, protection, mitigation, preparedness, response, and recovery efforts of post event consequences. The ERP is structured according to the Comprehensive Preparedness Guide (CPG) 101, Nov 2010, while also following the principles of the National Incident Management System (NIMS) and Incident Command System (ICS).

Furthermore, the ERP utilizes the Homeland Security Exercise and Evaluation Program (HSEEP) to addresses response, training, exercises, equipment, evaluation, and corrective action practices.

The (add school position) shall oversee or coordinate with applicable partners the following ERP actions:

- The ERP shall be reviewed annually and modified as necessary by (add position/committee).
- The school ERP shall be coordinated with the District and external agencies that may be affected by ERP implementation, in an effort to ensure consistency and compatible of their jurisdictional plans.
- Substantive changes between review periods, such as changes in roles or responsibilities, will prompt notification to listed stakeholders. Minor edits such as grammar or spelling changes will require no notification.
- If the organization and upkeep of the ERP includes process changes such as an ERP review, or a promulgation document designed to capture signature acknowledgement from each partner agency named within the document, the (add position/committee) will generate a draft document that will be sent to the relevant partners for review and recommendations.
- After a review period and consideration of stakeholder comments, the document will be finalized and signatures obtained.
- Final results of the reviews and any changes to the ERP shall be presented to the (add school position) for final approval before being adopted.
- Each school unit or department identified as having a role in this ERP is responsible for communicating the content of the ERP to their staff and ensuring key staff has the opportunity to attend ERP training and exercise activities.
- Ensure ERP compliance with the applicable local, State, and federal procedures.

B. Testing, Training, and Exercise

The development of the ERP Training and Exercise Plan is a key component of the School ability to respond to an emergency situation. It is imperative that all school staff have a general understanding of what (add School/District name) role will be during an event and the expected response protocol, which is structured by NIMS while also following the HSEEP procedures. Therefore, training and exercise opportunities will provide the required background and understanding of staff and response volunteers.

- ERP training opportunities, as well as review of preparedness or response support materials, shall be incorporated into the annual Training and Exercise schedule and Workforce Development Plan.

- Each school unit or department identified as having a role in this ERP is responsible for communicating the content of the ERP to their staff and ensuring key staff has the opportunity to attend and participate in ERP training and exercise activities.
- Working with response agency partners, HSEEP compliant exercises should be held to train school and response personnel and evaluate the adequacy of the ERP. Following HSEEP procedures, an After Action Report (AAR) and the Improvement Plan (IP) for each exercise shall be developed and documented appropriately.

C. Plan Contact Information

Name and Position	Phone Number	Alternant Phone Number
E-mail:		
Department:		

D. Record of Change

Change Number	Date of Change	Description of Change	Made By

E. Record of Distribution

Date	Version	Name and Title	Department/Agency

IX. AUTHORITIES AND REFERENCES

Authorities and References

This section provides the legal basis for emergency operations and activities. This section of the plan includes:

Lists of laws, statutes, ordinances, executive orders, regulations, and formal agreements relevant to emergencies (e.g., MAAs)

Specification of the extent and limits of the emergency authorities granted to the senior official, including the conditions under which these authorities become effective and when they would be terminated

Pre-delegation of emergency authorities (i.e., enabling measures sufficient to ensure that specific emergency-related authorities can be exercised by the elected or appointed leadership or their designated successors)

Provisions for COOP and COG (e.g., the succession of decision-making authority and operational control) to ensure that critical emergency functions can be performed

Procedures within this document apply to (add School/District name). The organizational and operational concepts set forth in these procedures are promulgated under the following:

LOCAL
(Insert any applicable local citations)
STATE
State Revised Statutes
FEDERAL

- Robert T. Stafford Disaster Relief and Emergency Assistance Act, PL 100-707
- Emergency Management and Assistance, Code of Federal Regulations, Title 44
- Superfund Amendments and Reauthorization Act of 1986, PL 99-499 (Title III, "Emergency Planning and Community Right-to-Know Act of 1986")
- Comprehensive Environment Response Compensation and Liability Act of 1980, PL 96510 (CERCLA or "Superfund")
- County Health Department Security and Bioterrorism Preparedness and Response Act (42 CFR Part 73)
- Homeland Security Act of 2002 (CIKR, Intro-2, CPG 101)
- Homeland Security Presidential Directive (HSPD) 3, 5 and 8:
- National Response Framework
- National Incident Management Systems (NIMS)
- Occupational Safety and Health Administration (OSHA) Rule 1910.120

X. DEFINITIONS

1. Common Terminology: Using common terminology helps to define organizational functions, incident facilities, resource descriptions, and position titles.
2. Demographic profile: Marketers typically combine several variables to define a demographic profile. A demographic profile (often shortened to "a demographic") provides enough information about the typical member of this group to create a mental picture of this hypothetical aggregate.
3. Disaster Recovery Center (DRC). The Disaster Recovery Center is established by FEMA in partnership with state and local emergency management offices. Representatives from federal, state, local, and volunteer agencies are there to explain the assistance available and to assist victims in procuring it.
4. Emergency Alert System (EAS). A network of broadcast stations and interconnecting facilities which have been authorized by the Federal Communications Commission to operate in a controlled manner during a war, state of public peril or disaster, or other national emergency—as provided by the emergency broadcast system plan. Supersedes EBS (Emergency Broadcast System).
5. Emergency Management (EM). A framework for organizing and managing emergency protection efforts. Prevention, protection, mitigation, preparedness, response, and recovery—in the all-hazards approach.

6. Emergency Operations Center (EOC). Specially equipped facilities from which government officials exercise direction and control and coordinate necessary resources in an emergency.
7. Emergency Public Information (EPI). Information that is disseminated to the public via the news media before, during, and/or after an emergency or disaster.
8. Emergency Response Plan (ERP). Contains policies and procedures to follow before, during and after an emergency. The ERP integrates emergency preparedness activities into one document. It is the focal point for School planning and preparedness procedures.
9. Emergency Situation. As used in this plan, this term is intended to describe a range of situations, from an incident to a major disaster. It includes the following:
    - Incident. An incident is a situation that is limited in scope and potential effects.
    - Emergency. An emergency is a situation that is larger in scope and more severe in terms of actual or potential effects than an incident.
    - Disaster. A disaster involves the occurrence or threat of significant casualties and/or widespread property damage that is beyond the capability of the local government to handle with available local resources.
10. Federal Emergency Management Agency (FEMA). The federal agency charged with development of an integrated emergency management system and with supporting emergency management and disaster assistance efforts at all levels of government. See: http://www.fema.gov.
11. Hazard: Something that is potentially dangerous or harmful, often the root cause of an unwanted outcome.
    - Human-Caused Hazard: A hazard that arises from deliberate, intentional human actions to threaten or harm the well-being of others. Examples include school violence, terrorist acts, or sabotage.
    - Natural Hazard: A hazard related to weather patterns and/or physical characteristics of an area. Often natural hazards occur repeatedly in the same geographical locations.
    - Technological Hazard: A hazard originating from technological or industrial accidents, infrastructure failures, or certain human activities. These hazards may cause loss of life or injury, property damage, social and economic disruption, or environmental degradation, and often come with little to no warning.
12. Hazardous Material (HAZMAT). A substance in a quantity or form posing an unreasonable risk to health, safety and/or property when manufactured, stored or transported. The substance, by its nature, containment and reactivity, has the capability for inflicting harm during an accidental occurrence. It may be toxic, corrosive, flammable, reactive, an irritant, a strong sensitizer and poses a threat to health and the environment when improperly managed. Included are toxic substances, certain infectious agents, radiological materials and other related materials such as oil or other petroleum products, and industrial solid waste substances.
13. Incident Commander (IC). The person responsible for the management of all incident operations. The IC is in charge of the incident site.

14. Incident Command Post (ICP). The location from which the Incident Commander oversees all incident operations. The ICP may be located outside, in a vehicle, trailer, or tent, or within a building. The ICP will be positioned at a safe distance from an accident site where the incident commander, responders and technical representatives can make response decisions, deploy man power and equipment, maintain liaison with the media and handle communications.
15. Incident Command System (ICS). The combination of facilities, equipment, personnel, procedures, and communications operating with a common organizational structure, with responsibility for the management of assigned resources to effectively accomplish stated objectives pertaining to an incident and/or event.
16. Continuity of Operations Plan (COOP) establishes guidance and procedures to ensure the resumption of essential functions in the event that an emergency or disruption incapacitates operations and/or requires the relocation of selected personnel and functions. (See Annex A COOP.)
17. Local Emergency Planning Committee (LEPC). A group of representatives of government and private industry who coordinate response plans for emergency conditions.
18. Lockdown. A procedure of locking classroom doors, covering windows, moving all persons away from windows and doors during a situation involving dangerous intruders, or other incidents that may result in harm to persons inside the school building.
19. Liaison Officer. A member of the command staff responsible for interacting with representatives from cooperating and assisting agencies.
20. Logistics Section. The section responsible for providing facilities, services, and materials for the incident.
21. Material Safety Data Sheet (MSDS). Document containing specific information on the safe handling of chemicals in the workplace.
22. National Weather Service (NWS). To provide weather and flood warnings, public forecasts and advisories for all of the United States, its territories, adjacent waters and ocean areas, primarily for the protection of life and property. NWS data and products are provided to private meteorologists for the provision of all specialized services. See: http://www.nws.noaa.gov.
23. NIMS provides a consistent framework for incident management at all jurisdictional levels regardless of the cause, size, or complexity of the incident. NIMS is not an operational incident management or resource allocation plan. NIMS represents a core set of doctrine, concepts, principles, terminology, and organizational processes that enables effective, efficient, and collaborative incident management.
24. Public Information Officer (PIO). A member of the command staff responsible for interfacing with the media or other appropriate agencies requiring information directly from the incident.
25. Radio Amateur Civil Emergency Service (RACES). A radio communication service conducted by volunteer licensed amateur radio operators, for providing emergency radio communications to local, regional, or state emergency management organizations. FCC 97.163(a).

26. Resources List. A current list of all resources (equipment, personnel, supplies), which can be used by emergency services in response to local disaster/emergencies.
27. Shelter-In-Place. A procedure addressing the need to provide refuge for students, staff and visitors within the school building during an emergency.
28. Staging Area (SA). A pre-selected location having large parking areas and cover for equipment, vehicle operators, and other personnel such as a major shopping area, schools, etc. The SA provides a base for coordinated emergency operations, assembly of persons to be moved by public transportation to reception jurisdictions, a rally point for mutual aid, or a debarking area for returning evacuees.
29. Transfer of Command: The process of moving the responsibility for incident command from one Incident Commander to another is called "transfer of command." It should be recognized that transition of command on an expanding incident is to be expected. It does not reflect on the competency of the current Incident Commander.
30. Unified Command. In ICS, Unified Command is a unified team effort that allows all agencies with responsibility for the incident, either geographical or functional, to manage an incident by establishing a common set of incident objectives and strategies. This is accomplished without losing or abdicating agency authority, responsibility, or accountability. The operations section chief is responsible for implementing the incident action plan.
31. Unity of Command and Chain of Command: Chain of command refers to the orderly line of authority within the ranks of the incident management organization. Unity of command means that every individual has a designated supervisor to whom he or she reports at the scene of the incident.

    These principles clarify reporting relationships and eliminate the confusion caused by multiple, conflicting directives. Incident managers at all levels must be able to control the actions of all personnel under their supervision.
32. Unified Command: In incidents involving multiple jurisdictions, a single jurisdiction with multiagency involvement, or multiple jurisdictions with multiagency involvement, Unified Command allows agencies with different legal, geographic, and functional authorities and responsibilities to work together effectively without affecting individual agency authority, responsibility, or accountability.

EMERGENCY RESPONSE PLAN TEMPLATE

ATTACHMENT—1
INFORMATION COLLECTION, ANALYSIS, AND DISSEMINATION

1. BACKGROUND INFORMATION
2. INFORMATION SHEET (Click Link to Open)
3. WORKSHEET (Click Link to Open)

1. Background Information

The goal of the Information Collection, Analysis, and Dissemination section is to gather critical information to enhance life safety, in addition to the physical resistance of our school and district to man-made and natural hazards.

Furthermore, we hope to achieve integrated infrastructure protection and disaster management, and at the same time meet the needs of the students, staff, administration, and public.

Protecting a school building and grounds from physical attack is a significant challenge because the design, construction, renovation, operation, and maintenance of a facility must consider numerous building users, infrastructure systems, and building design codes and in collecting, analyzing and disseminating this information we hope to assist you in capturing critical information that will assist school or district staff, and first responders when responding to and recovering from an emergency or disaster occurring on school property.

Schools are an integral part of every community in the United States. As of fall 2010, approximately 75.9 million people were projected as enrolled in public and private schools at all levels including elementary, secondary, and postsecondary degree-granting. In addition, the number of professional, administrative, and support staff employed in educational institutions was projected at 5.4 million (U.S. Department of Education 2010).

Additionally, schools serve as resources for their communities. Many schools are used as shelters, command centers, or meeting places in times of crisis. Schools are also used widely for polling and voting functions. In some communities, schools are places of health care delivery. Consequently, ensuring the safety of students, faculty, and staff in our schools, as well as the safety of the school buildings themselves, is critically important.

Schools may or may not be the targets, but could be indirectly threatened by collateral damage from an event occurring at nearby facilities.

Protecting a school against terrorist attack or active shooter is a challenging task. A school may have considerable vulnerabilities, because of its well-defined periods of use, designated access points, storage of sensitive personal information, minimal security forces, and numerous avenues of penetration and escape for attackers.

More in-depth information on this topic may found in the following materials:

- FEMA-428/BIPS-07/January 2012 Edition 2, this primer focuses on a single facility type with a very specific occupancy and vulnerability.
- FEMA P-424, Design Guide for Improving School Safety in Earthquakes, Floods, and High Winds (2010). In dealing with the protection of schools from school shooting or terrorist threats.
- BIPS 06 (Formerly FEMA 426), Reference Manual to Mitigate Potential Terrorist Attacks Against Buildings. BIPS 06 deals with all building types and occupancies.

17 State or Local Emergency Response Plan Template

The Basic Plan needs to be developed in coordination with other state agencies, nongovernmental organizations (NGOs), and private sector partners (PSPs) and continues to align with the National Incident Management System (NIMS), as well as the Department of Homeland Security (DHS) National Response Framework (NRF) and the National Disaster Recovery Framework (NDRF). In addition, the Basic Plan and its appendices, Emergency Support Function (ESF) Annexes, and Support and Hazard Specific Annexes should incorporate lessons learned from exercises, training, incidents, and events.

EMERGENCY MANAGEMENT PROGRAM

The strategies and objectives of the emergency management program are established in several plans, including the following:

- Secure Strategic Plan sets the overall course and direction of Preparedness, including the emergency management program, by defining its vision, mission, goals, and objectives.
- Emergency Operations Plan supports and includes annexes and is an all-discipline, all-hazards plan that establishes a single, comprehensive framework for the management of statewide events.
- The Critical Infrastructure Protection and Resiliency Strategic Plan supports the National Infrastructure Protection Plan (NIPP) by establishing a coordinated approach to national priorities, goals, and requirements for critical infrastructure and key resource protection.
- The strategic plan requires the development of sector-specific plans (SSPs) to provide the means by which the NIPP is implemented across all critical infrastructure and key resources sectors.
- The Hazard Mitigation Plan identifies hazards and analyzes the risk and potential impacts. The plan focuses on reducing the long-term vulnerability to identified hazards by establishing interim and long-term goals, objectives, and actions.
- The Recovery and Mitigation Standard Operating Guide describes the framework and associated actions to be taken in the establishment, operations, and demobilization of a joint field office and area coordinating officers, to deliver disaster assistance and support throughout all phases of recovery in a timely and effective manner.

- The Plan for the Public Assistance Grant Program identifies the roles, responsibilities, processes, and procedures for administering the Federal Emergency Management Agency (FEMA) Public Assistance program.
- Other hazard-specific plans developed by individual agencies to address specific incidents or pursuant to federal guidance.
- Agency strategic plans that focus on prioritized actions, including the functions of each agency, which are critical to the emergency response and recovery operations.
- Agency continuity plans that address an agency's ability to continue to provide essential government functions in the event of a disruption. Plans include orders of succession, delegations of authority, and essential records, systems, and equipment. They also address the procedures for restoring essential government functions, including those that are critical to emergency response and recovery operations.

COMPONENTS OF THE EMERGENCY OPERATIONS PLAN

The Basic Plan uses an all-hazards approach to incident management. It describes the concepts and structures of response and recovery operations, identifies agencies and coordinating NGOs and PSPs with lead and support emergency management functions, and defines emergency preparedness, response, recovery, and mitigation responsibilities of local governments. There are 17 ESF Annexes, 5 Support Annexes, and 7 Hazard-Specific Annexes to the Basic Plan.

ESF Annexes provide the structure for state coordinated emergency operations in support of affected local governments, individuals, and businesses. The annexes identify lead and supporting agencies, NGOs, and PSPs and explain in general terms how the local government will organize and implement support functions. Agencies, NGOs, or PSPs are assigned to lead or support the ESFs based on authorities, resources, and capabilities.

Support Annexes address those functions that may be applicable to every type of incident and provide support for all ESFs. They describe the framework through which state agencies, NGOs, and PSPs coordinate and execute the common functional processes and administrative requirements necessary to ensure efficient and effective incident management.

Hazard-Specific Annexes address contingency or hazard situations requiring specialized response and recovery procedures. They describe policies, situations, concepts of operations and responsibilities pertinent to incidents such as radiological emergencies, hurricanes, public health threats like pandemic influenza, terrorism incidents, technological hazards, and large-scale hazardous-materials incidents.

COMPONENT RESPONSE DIAGRAM

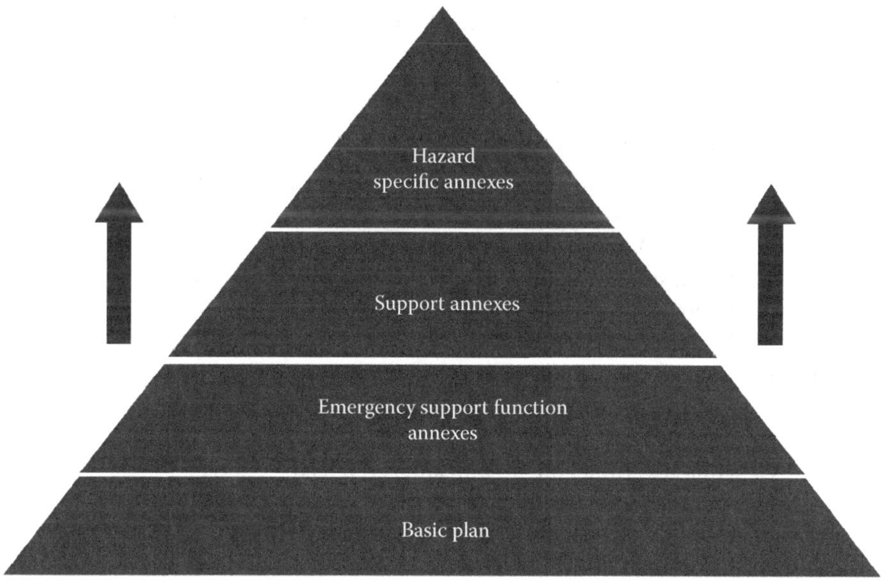

FIGURE 17.1 Plan flow from basic to complex.

The diagram in Figure 17.1 demonstrates the process in which the four components of the operation plan are activated. The Basic Plan, forming the foundation, is always activated during an incident.

As the state's response grows, select ESF annexes are called upon to support specific missions. Support annexes may then be implemented to supplement actions carried out under the Basic Plan and ESFs. Located at the top of the pyramid are Hazard-Specific Annexes, which play an important role during the state's response to particular incidents. These comprehensive, incident-specific documents address enhanced response actions not identified in the Basic Plan and underlying ESFs.

TABLE OF CONTENTS

13. Public Safety & Security
14. Recovery & Mitigation
15. External Affairs
16. Military Affairs
17. Volunteer & Donations Management

Support Annexes

1. Continuity of Government
2. Recovery Programs
3. Finance & Administration
4. Mass Care & Sheltering
5. Evacuation & Re-entry

Hazard Specific Annexes

1. Radiological Emergency Response
2. Terrorism Consequence Management (restricted)
3. Hurricane & Tropical Storm Response
4. Pandemic Influenza Response
5. Hazardous Materials Response
6. Technological Hazards Response (restricted)
7. Earthquake Response

I. INTRODUCTION

The plan provides the coordinating structures, processes, and mechanisms in a collective framework for the coordination of state support to affected local governments, individuals, and businesses. It is compatible with the NRF and NDRF, and provides the structure for coordinating with the federal government in the delivery of federal disaster assistance. The plan supports the emergency management mission of the Commonwealth by improving the capability to respond to and recover from natural and human-caused disasters.

A. Purpose

The Basic Plan is the foundation of the local/state's emergency response and recovery organization. This plan identifies the role of state government and coordinating NGOs and PSPs before, during, and after a disaster, large-scale emergency, or event affecting the local government. It establishes the concepts and policies under which all elements of state government and coordinating NGOs and PSPs will operate during emergencies. It also provides the framework within which more detailed emergency plans or procedures can be developed and maintained by state agencies, local governments, NGOs, and PSPs.

To ensure the local government capability to implement this plan, each agency tasked with essential emergency management responsibilities shall develop and maintain response plans in support of the plan. In addition, executive branch agencies and institutions of higher education are required to develop and maintain an all hazards continuity plan that identifies the critical and time-sensitive functions, applications, and processes of the agency to be recovered and continued in an emergency or disaster, including alternate operating capabilities. Agency functions critical to the state emergency response and recovery operations shall have restoration priority.

Top priorities for incident management are to:

- Lives and protect the health and safety of the public, responders, and recovery workers.
- Protect property and mitigate damages and impacts to individuals, communities, and the environment.
- Protect and restore critical infrastructure and key resources. Facilitate recovery of individuals, businesses, communities, governments, and the environment.
- Manage public expectations regarding response and recovery activities.

B. Scope and Applicability

The plan covers the full range of complex and constantly changing requirements in anticipation of or in response to threats or acts of terrorism, major disasters, other emergencies, and events within or affecting the local government.

The plan establishes interagency, multi-jurisdictional, and public/private mechanisms for state agency involvement in domestic incident management operations. These mechanisms include coordinating structures and processes for incidents requiring:

- Local-to-local support through Statewide Mutual Aid (SMA).
- State support to cities, counties, and the towns.
- State-to-state support through Emergency Management Assistance Compact (EMAC) agreements or other appropriate instruments.
- Public and private sector incident management integration.

This plan is applicable to all agencies of the local government, and coordinating NGOs and PSPs that may be requested to provide assistance or conduct operations in response to an incident or event. Depending on the severity of the incident or event, the governor may declare an emergency or determine that state emergency management coordination is required in order to protect or save lives, prevent or minimize property or environmental impacts, or otherwise assist affected areas. Moreover, this plan also provides the foundation for the organization and coordination of community recovery and mitigation activities.

C. Situation and Planning Assumptions

This plan was developed with the following assumptions:

1. Local/State agencies assigned responsibilities in the plan have developed and will maintain the necessary plans, standard operating procedures, mutual aid agreements, and model contracts to successfully accomplish their tasks.

2. Local/State agencies are prepared to fulfill responsibilities assigned to them in the plan, supporting plans, and joint operational or regional plans.
3. Local/State agencies' capabilities to carry out response and recovery tasks are enhanced through the development, maintenance, and testing of agency continuity plans.
4. Coordinating NGOs and PSPs have internal plans or procedures specific to their assigned roles and responsibilities outlined in the plan.
5. Incidents, including large scale emergencies or events, require full coordination of operations and resources, and may:
   - Occur at any time with little or no warning.
   - Require significant information sharing across multiple jurisdictions and between the public and private sectors.
   - Involve single or multiple jurisdictions and/or geographic areas.
   - Have significant statewide and/or national impact and/or require significant inter-governmental coordination.
   - Involve multiple, highly varied hazards or threats on a local, regional, statewide or national scale.
   - Result in mass casualties; displaced persons; property loss; environmental damage; and disruption of the economy and normal life support systems, essential public services, and basic infrastructure. These effects may be minimized by the proactive notification and deployment of state resources in anticipation of or in response to major incidents in coordination and collaboration with local, private and federal entities.
   - Require resources to assist individuals with access and functional needs.
   - Impact critical infrastructures across sectors.
   - Exceed the capabilities of state agencies, local governments, NGOs, and PSPs (to include private sector infrastructure owners and operators).
   - Attract a sizeable influx of public, private, and voluntary resources, including independent and spontaneous volunteers.
   - Require short-notice state asset coordination and response.
   - Require prolonged, sustained incident management operations and support activities for long-term community recovery and mitigation.
6. The factors influencing the need for federal involvement in response and recovery may include, but are not limited to:
   Severity and magnitude of the incident.
   - State or local needs exceeding available resources. Incident type or location. The need to protect the public health, welfare, or the environment.
   - The economic ability of the state and/or the affected localities to recover from the incident.
7. The combined expertise and capabilities of government at all levels, PSPs, and NGOs will be required to prepare for, respond to, mitigate, and recover from incidents of major or catastrophic proportions.
8. Localities have a plan as part of an emergency management program that reflects current doctrine and protocol, and includes provisions for the needs and requirements of the population, such as children, individuals with

disabilities or access and functional needs, service animals, and household pets.

9. The negative effects on residents and its economy from natural disasters such as hurricanes, floods, winter storms, earthquakes, and wildfires may increase due to increased urban development in vulnerable areas, industrial expansion, traffic congestion and widespread use and transport of hazardous materials. These factors may also increase the risk of human-caused emergencies such as hazardous materials accidents, gas pipeline accidents, power failures, resource shortages and environmental contamination. In addition, depending on the proximity to any federal facilities located throughout area increase the potential for terrorism related incidents.

D. Policies

1. The Emergency Services and Disaster Laws require the state, and each city, county, and certain towns to establish an emergency management organization, and develop and maintain a current EOP structured around the existing constitutional government.
2. Incidents are typically managed at the lowest possible geographic, organizational, and jurisdictional level. As such, direction of emergency operations will be exercised by the level of affected local government closest to the incident. If an incident is determined to have a terrorism nexus, appropriate federal agencies will likely assume direction and control of the investigation, in coordination with state and local authorities.
3. Coordination of emergency response will be accomplished within the Incident Command System, allowing for the incorporation of local, state, and federal agencies, and other responsible parties.
4. The provision of state assistance and the deployment of resources for emergency response are dependent upon the receipt of prompt and accurate situational information from local officials. Local requests for assistance and/or situation reports will be submitted utilizing WebEOC. If WebEOC is unavailable, information may be transmitted via other means, such as fax, email, telephone, or radio.
5. The Governor may issue an Executive Order declaring that a "state of emergency" exists in all or a portion of the state.
6. All Executive Branch agencies (including state institutions of higher education) are resources of the Governor, and therefore may be activated to support the emergency response plan during response or recovery activities.
7. Requests for assistance from entities including, but not necessarily limited to, nursing homes, colleges and universities, and authorities will be submitted to the local Emergency Management Coordinator in the jurisdiction in which the entity resides.

   Requests for assistance will be submitted to the local Emergency Management Coordinator or their designee when local capabilities are exceeded.

8. Emergency response and recovery activities will be initiated and conducted using the principles contained in NIMS.
9. Assistance from state agencies will be coordinated by the state EOC.
10. Resources furnished to local governments from state agencies, NGOs, and PSPs will be assigned on a mission-type basis to the appropriate ESF and will be under the operational control of the agency, or organization furnishing the resources. These resources will be furnished as soon as reasonably possible.
11. If the need for resources exceeds the capabilities of localities and the state, the Governor may request federal assistance. The Governor may request the President declare an emergency or major disaster, and may request assistance for individuals, public entities, state agencies, and local governments.
12. During incidents for which the President has declared an emergency or major disaster, federal support is delivered in accordance with relevant provisions of the Robert T. Stafford Act, as well as other available disaster assistance programs in coordination with the appropriate agencies of the state.

II. HAZARD IDENTIFICATION AND RISK ASSESSMENT

Preparedness, response, recovery, prevention, and mitigation strategies are largely based on analyses of natural and human-caused hazards with a history of or potential for impacts. The Hazard Mitigation Plan identifies and assesses risk to both natural and human-caused hazards. The hazards are categorized based on their characteristics: natural, technological, hazardous materials, and human-caused. In addition, a threat hazard identification and risk assessment (THIRA) has been completed for the Commonwealth, which addresses capability gaps for specific high consequence hazards. The plan is designed to address all hazards identified.

III. ORGANIZATIONAL STRUCTURE

The state provides that emergency management organizations and operations will be structured around existing constitutional government. The Director of Emergency Management for the state appoints a Coordinator of Emergency Management (State Coordinator). The State Coordinator appoints the managers and coordinates response activities. The State Coordinator also appoints the Director of Recovery and Mitigation, who coordinates state recovery activities for a disaster that is declared by the President as authorized by the Stafford Act. For a presidentially declared disaster, the State Coordinator is usually designated the State Coordinating Officer by the Governor.

The planning and coordination of state, local, and federal entities to accomplish productive and timely recovery efforts. All agencies, NGOs, and PSPs will coordinate with state coordinator to develop and maintain their designated part(s) of the plan.

The organization for state level emergency operations includes:

The Governor and his immediate staff.

- State resources coordinated by emergency manager during routine operations and by agencies during augmentations. Prior to a declaration, the Governor (with or without the recommendation of the State Coordinator), may issue a deployment authorization of National Guard and/or Defense Force personnel. The State Coordinator will be notified of authorization approvals, including any limitations other than those specified in code, either directly from the Governor or through the Secretary of Public Safety and Homeland Security.
- State agencies assigned emergency responsibilities in the plan. This plan identifies ESFs and designates responsibilities to lead and support agencies for each function.
- Coordinating NGOs designated in the plan and/or affiliated with ESFs. Some NGOs enter into agreements with the state to perform specific disaster functions.
- PSPs designated in the plan and/or working with the lead state agency for each ESF. Additional PSPs may be incorporated into the state response as deemed necessary by the state Coordinator.
- Recovery activities coordinated by state emergency manager for the short and interim phases and by the Economic Crisis Strike Force (ECSF) for interim and long-term phases is to respond as needed to economic disasters in the local communities by assisting localities in developing short and long-term strategies to address the crisis and provide a single point of contact for citizens in affected communities. This Strike Force is led by the Secretary of Commerce and Trade, in coordination with other Secretariats as appropriate.
- Local governments: Each city and county, as well as certain towns, are required to have an emergency management organization as defined in the state code. Locally available manpower, materials, equipment, and facilities should be identified in each local EOP. Localities not affected may be asked to provide assistance through the mutual agreements Program.
- Federal Emergency Management Agency (FEMA) and VDEM coordinate the delivery of disaster assistance from a Joint Field Office (JFO). The state organization for recovery and mitigation parallels the federal organization at the JFO to ensure the effective interface and coordination from both a functional and programmatic perspective. The recovery framework is structured to be flexible and scalable to quickly adapt to any disaster situation.
- Other Federal agencies when requested and within their statutory authority.

IV. CONCEPT OF OPERATIONS

This section describes the state emergency management coordinating structures, processes, and protocols employed for incident management. These coordinating structures and processes are designed to enable execution of the responsibilities of the Governor through the appropriate state agencies. It integrates federal, state, local, NGOs and PSPs efforts into a comprehensive statewide approach to incident management.

The state plan and local emergency operations plans are founded upon the concept that emergency operations begin at the jurisdiction level closest to the event, and are managed by local police, fire, emergency medical and health, emergency management, and other response personnel. In the vast majority of disasters, local resources will operate under the umbrella of a mutual aid agreement or compact to provide the first line of emergency response and incident management support. In some instances, a state or federal agency with principal activities or facilities in a local area may act as a first responder and may provide direction or assistance consistent with its specific statutory authorities and responsibilities. State assistance will be provided upon request when needs exceed local capabilities.

Situations in which several localities are threatened or affected concurrently may involve the state from the onset. If the state capabilities are exceeded, the Governor may request federal assistance. At each level, the government should officially declare a 'state of emergency' to exist in order to request assistance. A local emergency declaration indicates that local resources capable of handling the situation are fully committed before state assistance is requested. Likewise, state resources should be fully committed before federal assistance is requested. Exceptions are made to these procedures when localities request state assistance with search and rescue or hazardous materials response.

The EOC has four operational levels:

1. Routine Operations: Emergency Operations plans and procedures are developed and maintained. Training and exercises are conducted periodically as required to maintain readiness.
2. Increased Readiness: When a disaster threatens, all agencies, and coordinating NGOs, and PSPs assigned responsibilities in the COVEOP will take action as called for in their respective parts of the plan.
3. Response Operations: Full-scale operations and a total commitment of manpower and resources are required to mobilize and respond in time of emergency.
4. Recovery Operations: Essential facilities and services are restored. Displaced persons return to their homes. Federal disaster assistance programs are implemented. Severely damaged structures are rebuilt, or demolished and replaced.

Damaged transportation, water, and communication infrastructure are replaced. Normal conditions and the economy are restored. The duration of this period may extend for two or more years, depending upon the severity of the disaster.

The operational levels are used for all emergencies and are not related to the emergency classification levels for fixed nuclear facilities referenced in the Radiological Emergency Response Annex of the COVEOP, or the Readiness Condition Action Guidelines referenced in the Hurricane and Tropical Storm Response Annex of the plan.

In time of emergency:

- Provides centralized state government emergency operations in order to coordinate the delivery of resources to local governments.
- Provides advice and counsel to the Governor or the Governor's designee to formulate policy.
- Establishes priorities.
- Collects, analyzes, and disseminates situational awareness information.
- Provides sufficient staff to maintain communications with the federal, state, local, and private sector partners related to conditions or developing situations related to the emergency.

Incident Management Actions

Incident management begins with identification of a risk, threat, or actual event, and concludes when those affected have been restored to pre-incident conditions, when feasible. Actions may include, but are not limited to, notification and warning, activation of the response actions, assessment of impact, request for federal assistance, and recovery and mitigation activities.

A. Notification

State agencies, local governments, NGOs, and PSPs report threats, incidents, and potential incidents using established communications and reporting channels.

The Fusion Center receive threat and operational information regarding incidents or potential incidents, and then makes an initial determination to implement Federal information sharing and incident management protocols.

B. Alert and Warning

Public notifications shall be disseminated through multiple means that may include:

- Public Announcements
- Social Media
- Watches and Warnings
- Emergency Alert System (EAS)
- Other Emergency Bulletins

C. Activation of the EOC/Emergency Response Team (ERT)

The EOC receives notification of conditions, events, and/or occurrences that may affect the state, reports of incidents or hazardous conditions existing within the state, and requests for assistance from local governments.

In support of the EOC's mission, is to maintain an Incident Management Cadre (IMC) to assist the State Coordinator in determining the requirement to augment the ERT. The IMC consists of the VERT Coordinator, the Staff Duty Officer, Operations Chief, Plans Section Chief, Logistics Chief, Finance and Administration Chief, Recovery Chief, External Affairs Chief, Search and Rescue Duty Officer, and Communications Maintenance Officer. Subject matter experts may be requested to participate as needed.

The ERT Coordinator will, based on information provided from the IMC, brief the State Coordinator and/or his or her designee(s) of the recommended operational course of action.

The ERT Coordinator, in consultation with the State Coordinator, Deputy State Coordinator, and/or his or her designee, will initiate notification of the appropriate VERT Staff and necessary state agency, NGO, and PSP points of contact of an augmentation and request representatives to support the ERT.

Based on an analysis of the information received, the State Coordinator may recommend to the Governor that he/she declare a state of emergency, so that all necessary state resources can be immediately prepared or deployed to affected areas.

D. Response Actions

At the request of local governments, the ERT directs state response activities that address the short-term direct effects of an incident. These activities include immediate actions to preserve life, property, and the environment, meet basic human needs, and maintain the social, economic, and political structure of the affected community.

As part of an effective response, the ERT must continuously refine its ability to assess the situation as an incident unfolds and rapidly provide accurate and accessible information to decision makers. This is accomplished by maintaining situational awareness and a common operating picture through receipt of prompt and accurate information from affected localities using established reporting mechanisms.

Response actions may include the following resources and/or elements:

- Law enforcement
- Fire and emergency medical services
- Evacuation and transportation management
- Emergency public information or other actions taken to minimize additional damage
- Urban search and rescue services
- Deployment of advance teams to assess the severity of impact and expedite the delivery of state resources
- Establishment of mass care facilities, including shelters and feeding operations

- Establishment of a Family Assistance Center
- Provision of public health and medical services, food, water, and other emergency essentials
- Debris clearance and/or the emergency restoration of critical infrastructure
- Control, containment, and removal of environmental contamination

E. Assess Damage and Analyze Impact

An assessment of the damages and impacts of an incident begins at the local level. Local officials and state agency representatives are required to submit situation reports to the EOC throughout the course of the incident, and to send an Initial Damage Assessment (IDA) to the EOC within seventy-two (72) hours after the impact of the event.

The time frame may be adjusted by the ERT depending on the type of incident and/or circumstances. It is critical that the information provided in the IDA accurately reflect the damages and impacts sustained by the community, as the decision to request a Preliminary Damage Assessment and seek a Presidential disaster declaration is made based on this information.

As the situation changes and new information becomes available, local officials and state agency representatives will update the VERT as soon as reasonably possible. IDAs should be submitted to the EOC through WebEOC or on the standard forms provided by EM. ERT staff summarizes damage assessments from all sources and evaluates the impact of the incident on localities and the state. These assessments are used by the ERT to determine if state resources are sufficient to manage the incident.

Depending on the findings from the Initial Damage Assessment submitted by localities and state agencies, the State Coordinator may request FEMA conduct a joint Preliminary Damage Assessment (PDA) in coordination with affected localities or state agencies, to verify the damages and estimate the amount of supplemental assistance needed. ESF #14 will work with FEMA and other ESFs as appropriate, to coordinate joint PDAs. Joint PDAs focus on the more significantly affected areas identified on the Initial Damage Assessments submitted that may be eligible for federal assistance under the Robert T. Stafford Act, as amended (e.g., Individual Assistance, Public Assistance and Hazard Mitigation), or the Small Business Administration (SBA) Disaster Loan Program. State and federal personnel knowledgeable in these programs will accompany local or state officials to the damaged sites.

F. Request for Federal Assistance under the Stafford Act

The State Coordinator will report the findings of the PDAs to the Governor and make a recommendation regarding a request for federal assistance. Based on findings, the Governor will then send a letter to the President requesting a declaration of emergency or major disaster for specific localities in the states for Individual Assistance, Public Assistance, Hazard Mitigation Assistance, or all three.

When a large-scale disaster is imminent, it might be obvious that the scope of the event justifies federal assistance. For an expedited declaration, the Governor relies on forecasted or visual impact rather than damages verified through the PDA process.

When a Governor's request for a federal declaration is approved, the declaration will specify the federal assistance programs that will be made available to the state, as well as the jurisdictions included in the action. A Public Assistance Disaster Declaration provides the basis for financial and other forms of aid to state and local governments, and non-profit organizations for debris removal, implementation of protective measures, and damages sustained to critical infrastructure.

An Individual Assistance Disaster Declaration provides the basis for financial and other forms of aid to private citizens and, to a more limited extent, to businesses. When a disaster event does not meet the criteria for a Stafford Act Individual Assistance Disaster Declaration, it may meet the criteria for a Small Business Disaster Declaration. If requested and approved, hazard mitigation assistance may be available through the Hazard Mitigation Grant Program.

This program is based on 15% of the total federal assistance received through the Individual Assistance and/or Public Assistance programs. The Governor can request a SBA declaration if the findings of the SBA damage assessment process fulfills the criteria for a declaration. Other federal disaster assistance declarations can be requested independently including the U.S. Department of Agriculture and DHS/FEMA Fire Management Assistance.

G. Recovery Actions

Recovery involves actions needed to assist individuals and communities in returning to pre-incident conditions when feasible, as well as mitigating the potential impacts of future events. The state joins with federal agencies to establish a JFO that serves as the central coordination point among federal, state, local and voluntary organizations for delivering recovery assistance programs. ESF #14 and the ECSF will be the organizational framework used to plan, implement, and coordinate the incident specific short, interim, and long-term components of recovery. The ECSF will focus primarily on the long-term phase and be organized in the context of the Recovery Support Functions described in NDRF.

Recovery actions typically include direct financial assistance to eligible individuals who have lost residential dwellings or personal property, the coordination and execution of service and site restoration plans, and the reconstitution of state and local government operations and services through private sector, non-governmental, and public assistance programs.

The Hazard Mitigation Grant Program is a component of the Public Assistance Program, and is also implemented statewide, based on a strategy designed to mitigate future damage throughout the state.

Short-Term Recovery

The short-term component of the recovery process supports a variety of activities, including reviewing and assessing IDA's, recommending, arranging, and supporting PDAs, developing the request for a federal declaration, and if the declaration request is approved, establishing a JFO with FEMA. The JFO focuses primarily on the delivery of disaster assistance programs authorized by The Stafford Act, the management of the cost reimbursement process, and the coordination of available resources and support from local, state, federal, and non-governmental organizations, as well as the private sector.

Long-Term Recovery

The long-term recovery component of ESF #14 will assess the severity of impacts on the community, and begin defining the anticipated long-term recovery needs and resource requirements. The assessment associated with the long-term phase is broader in scope and focuses on the social and economic impact to the community and associated infrastructure.

Factors that are considered in this process include:

- Impacts on the housing sector and the projected number of displaced persons on a long-term basis.
- The severity and scope of the infrastructure damage precipitating major service and economic disruptions, and impacting negatively on overall response and recovery operations.
- Impacts on key sectors of the communities' economic base.

Based on the initial assessment of long-term recovery needs and resource requirements, and following consultation with the appropriate state officials and support agencies, the ESF #14 long-term recovery team will make a recommendation with regard to activating the ECSF. The ECSF will begin developing the necessary plans, priorities, initiatives and assistance to address the long-term recovery issues identified. The long-term recovery liaison team will continue to interface and coordinate with the short-term recovery component until the housing and associated issues of the shelter population have been addressed. The ECSF will support and be represented in the JFO as required and continue to support long-term recovery processes until objectives have been fulfilled.

State Public Assistance Program—Emergency Relief to Localities

For incidents that do not meet the level of impact to result in a federal declaration, state recovery programs may be implemented with or without a field office near the disaster site. The state codes established a state public assistance program entitled Emergency Relief to Localities to assist local governments in the recovery of eligible costs associated with localized major emergencies or disasters that lack sufficient damages to warrant a Presidential Declaration. The program, which is administered by DEM, is structured like the federal Public Assistance Program in terms of categories of work and having a threshold requirement. The program is considered one of last resort for those local jurisdictions that cannot meet the full cost. Any assistance provided under the program is at the discretion of the Governor.

H. Mitigation Actions

DEM maintains the state Hazard Mitigation Plan separate from the plan due to the federally mandated content and a separate review schedule. The goal of this plan is to reduce the impacts of hazards on human, economic, critical infrastructure, and natural resources throughout the state by incorporating mitigation concepts and objectives into existing and future policies, plans, regulations, and laws in the state, improving the quality of the data and analysis used in the hazard identification and risk assessment process, identifying and implementing projects that will eliminate long-term risk, directly reduce impacts from hazards, and maintain continuity of operations, and promoting awareness of hazards and potential mitigation strategies in order to increase resiliency.

Hazard mitigation involves reducing or eliminating long-term risk to people and property from hazards and their impacts. The JFO is the central coordination point among federal, state, local agencies, and NGOs for beginning the process that leads to the delivery of post disaster hazard mitigation assistance.

Mitigation measures may be implemented prior to, during, or after an incident, and in accordance with stated objectives and strategies in the states Hazard Mitigation Plan.

The joint federal-state mitigation unit in the JFO coordinates the delivery of mitigation programs within the affected area, including:

- Grant programs for loss reduction measures (if available).
- Delivery of loss reduction building-science expertise.
- Coordination of federal flood insurance operations and integration of mitigation with other program efforts.
- Conducting hazard recovery mapping to permit expedited and accurate implementation of both recovery and mitigation programs.
- Predictive modeling to protect critical assets.
- Early documentation of losses avoided due to previous hazard mitigation measures.
- Community education and outreach necessary to foster loss reduction.
- Education materials to affected property owners.

V. ROLES AND RESPONSIBILITIES

A. State Government—Governor of the State

As the chief executive officer of the state, the Governor is responsible for the public safety and welfare of the people of the state.

The powers and duties of the Governor in and or Emergency manager for emergency management are summarized below.

The Governor/Director Emergency Manager: (DEM)

- Is responsible for implementing the EMP and coordinating state resources to address the full spectrum of actions to respond to and recover from incidents in an all-hazards context to include terrorism, natural disasters, accidents, and other contingencies.

- If appropriate, makes a verbal declaration of a "state of emergency," to include an authorization to grant or seek temporary overweight, over width, registration, license, or hours worked exemptions to carriers transporting emergency relief supplies or providing utility restoration services. These verbal orders will be followed by a written authorization and an executive order.
- If appropriate, amends and rescinds orders and regulations and/or directs and compels evacuation of all or part of the populace from any threatened or affected area.
- Ensures the provision of essential services, including emergency and disaster response and recovery activities by executive branch agencies and institutions of higher education.
- Provides leadership and plays a key role in communicating to the public and in helping people, businesses, and organizations cope with the consequences of any type of declared emergency within any jurisdiction of the state.
- Encourages participation in mutual aid and implements authorities for the state to enter into mutual aid agreements with other states to facilitate resource sharing.
- May request federal assistance when it becomes clear that the capabilities of the state will be insufficient or have been exceeded or exhausted.
- May expend "sum sufficient" monies.
- May, without an Emergency Declaration, authorize the deployment of up to 300 members of the National Guard or Virginia Defense Force and materials (pursuant to state codes) to assist in times of emergency. Deployments under this code section are limited to a maximum of 300 personnel not to exceed five days, unless a state of emergency is declared.
- May provide financial assistance to localities.
- Serves as the Commander-in-Chief of the Commonwealth military forces pursuant to the state code.

State Coordinator and DEM

The State Coordinator of the Department of Emergency Management serves as the State Emergency Planning Director and has the following powers and duties as summarized below.

The State Coordinator and DEM:

- Implements portions of the plan to provide timely assistance to localities for non-declared incidents or events.
- Coordinates and provides guidance and assistance to affected political subdivisions to ensure orderly and timely response to and recovery from disaster effects.
- Coordinates disaster response actions of federal, state, and volunteer relief agencies.

- Establishes and maintains liaison with affected political subdivisions.
- In an emergency which does not warrant a gubernatorial declaration of a state of emergency, may after consultation with and approval of the Secretary of Public Safety and Homeland Security, enter into contracts and incur obligations necessary to prevent or alleviate damage, loss, hardship, or suffering caused by such emergency and to protect the health and safety of persons or property.
- Determines requirements for disaster relief and recovery assistance.
- Is responsible for ensuring that state response and recovery actions consider the resources needed for individuals with disabilities and/or access and functional needs.

Emergency Response Team (ERT)

The ERT consists of representatives from state agencies, NGOs, and PSPs who may be supported by federal partners. During a disaster, ERT personnel ensure that needed resources are provided to disaster stricken areas.

All ERT agencies and organizations have the responsibility to:

Maintain situational awareness through the collection, analysis, and dissemination of information and intelligence data.

- Receive, track, and coordinate requests for resources.
- Ensure all incident-applicable ESFs are staffed.
- Implement plans to coordinate emergency management efforts among local, state, and federal entities; as well as NGOs and PSPs.

Facilitate resolution of legal, policy, political, social, and economic concerns of the affected jurisdiction(s) as they affect response and recovery operations.

- Facilitate formulation of Protective Action Decisions, as needed.
- Facilitate demobilization plans and procedures.
- Participate in the after action report process.

State Agencies and Institutions of Higher Education

It is the responsibility of state government to provide for the well-being of the citizens of the state and ensure the continuity of state government operations, including the delivery of essential state government services during a disaster as tasked in this plan. Therefore, state agencies and institutions of higher education must continue to be prepared for all disasters.

All state agencies and institutions of higher education have the responsibility to:

- Have plans or procedures to implement their specific responsibilities outlined in the EOP.
- Have documented emergency action and continuity plans for their response to and recovery from a natural or human-caused disaster.

Direct agency area supervisors (from applicable division, district, or local offices) to participate in the local emergency planning process and become a part of the local emergency response organization, as appropriate.

- During small localized events, when possible, provide assistance directly to political subdivisions.

B. Local Government

The state code designates powers and duties for emergency management to political subdivisions (summarized below). Each jurisdiction:

- Shall have a director of emergency management.
- May request assistance from the state when local resources and capabilities are overwhelmed.
- The local Emergency Management Director in the political subdivision within the disaster area may, under the supervision and control of the Governor or his designated representative:
- Control, restrict, allocate or regulate the use, sale, production, and distribution of food, fuel, clothing, and other commodities, materials, goods, services and resource systems which fall within the boundaries of that jurisdiction and which do not impact systems affecting adjoining or other political subdivisions.
- Enter into contracts and incur obligations necessary to combat threatened or actual disaster, protect the health and safety of person and property, and provide emergency assistance to the victims of such disaster.
- Proceed without regard to time-consuming procedures and formalities prescribed by law (except mandatory constitutional requirements) pertaining to the performance of public work, entering into contracts, incurring of obligations, employment of temporary works, rental of equipment, purchase of supplies and materials, levying of taxes, and appropriation and expenditure of public funds.

In addition, the local jurisdiction:

- Shall identify a primary party responsible for managing emergencies within their jurisdiction.
- Is responsible for coordinating local resources to address the full spectrum of actions to respond to, and recover from incidents involving all hazards including terrorism, natural disasters, accidents, and other contingencies.
- Is responsible for ensuring that local actions include and accommodate individuals with disabilities and/or access and functional needs.

- May develop, or cause to be developed, mutual aid agreements for reciprocal assistance in the case of a disaster too great to be dealt with unassisted. Such arrangements shall be consistent with state plans and programs and it shall be the duty of each local organization for emergency management to render assistance in accordance with the provisions of such mutual aid agreements.
- Will receive and fulfill requests for assistance from entities including, but not necessarily limited to, nursing homes, colleges, universities, and authorities within their jurisdiction until local capabilities have been exceeded or exhausted.
- Will coordinate with state and federal officials after a disaster to implement recovery and mitigation strategies and programs.
- May form strong partnerships with citizen groups and organizations who provide support for incident management response, recovery, and mitigation. Local citizen groups, such as Community Emergency Response Teams and Medical Reserve Corps may coordinate with local and state agencies to provide resources to augment response and recovery activities.

C. Non-Governmental Organizations (NGOs)

NGOs (including voluntary organizations) may:

- If appropriate, provide specific disaster relief services during response and recovery in cooperation with state or local officials. This includes collaborating with first responders, governments at all levels, and other agencies and organizations providing relief services to sustain life, reduce physical and emotional distress, and promote recovery of disaster victims.

D. Private Sector Partners (PSPs)

Private sector organizations are critical to the capabilities of the ERT. They may:

- If appropriate, provide specific disaster relief services during response and recovery in cooperation with state or local officials.
- Support the ERT by participating in the planning process, sharing information, identifying risks, performing vulnerability assessments, developing emergency response and business continuity plans, enhancing their overall readiness, implementing appropriate prevention and protection programs, and donating or otherwise providing goods and services through contractual arrangement or government purchases to assist in response to and recovery from an incident.

Table 17.1 summarizes the roles of private sector organizations.

TABLE 17.1
Affected Organization or Infrastructure

Organization	Role
Affected Organization or Infrastructure	Private-sector organizations may be affected by direct or indirect consequences of the incident. These include privately owned critical infrastructure, key resources, and other private sector entities that are significant to local, regional, and national economic recovery from the incident. Examples of privately owned infrastructure include transportation, telecommunications, private utilities, financial institutions, and hospitals. Critical infrastructure and key resources are grouped into 18 sectors that together provide essential functions and services supporting various aspects of the state government, economy, and society. *Homeland Security Presidential Directive 7* establishes a national policy for Federal departments and agencies to identify and prioritize critical infrastructure and to protect them from terrorist attacks. The directive defines relevant terms and delivers 31 policy statements. These policy statements define what the directive covers and the roles various federal, state, and local agencies will play in carrying it out.
Response Resource	Private-sector entities provide response resources (donated or compensated) during an incident, including specialized teams, essential services, equipment, and advanced technologies through local public-private emergency plans or mutual aid and assistance agreements, or in response to requests from government and non-governmental volunteer initiatives.
Regulated and/or Responsible Party	Owners/operators of certain regulated facilities or hazardous operations may be legally responsible for preparing for and preventing incidents from occurring and responding to an incident once it occurs. For example, Federal regulations require owners/operators of nuclear power plants to maintain emergency plans and facilities, and to perform assessments, prompt notifications, and training for response to an incident.
State/Local Emergency Organization Member	Private-sector organizations may serve as an active partner in local and state emergency preparedness and response organizations, planning, and activities.
Components of the State Economy	As the key element of the state economy, private-sector resilience and continuity of operations planning, as well as recovery and restoration from an actual incident, represent essential emergency management activities.

VI. PLAN MANAGEMENT AND MAINTENANCE

A. Coordination

The state uses the "preparedness organization" concept described in the NIMS for preparedness and maintenance of the EOP. The organization includes all agencies, NGOs and PSPs with a role in state level incident management and provides a forum for coordination of planning, training, equipping, and other preparedness requirements.

The following entities are critical to the state's preparedness efforts to include the development and maintenance of this plan:

- The Secretary of Public Safety and Homeland Security (SPSHS)
  The DEM reports. SPSHS works with federal, state, and local officials, as well as the private sector, to develop a seamless, coordinated security and preparedness strategy for implementation of this plan and appropriate state-level response efforts.
- Secure state Panel
  Appointed by the governor and assigned the responsibility to monitor and assess the implementation of statewide prevention, preparedness, response and recovery initiatives...
- Department of Emergency Management (DEM)
  The state code assigns the Department of Emergency Management the responsibility to:
  - Prepare and maintain the State Emergency Operations Plan for disaster response and recovery operations that assigns primary and support responsibilities for basic emergency services functions to state agencies, organizations, and personnel as appropriate.
  - Coordinate and administer the disaster mitigation, preparedness, response, and recovery plans and programs with the proponent federal, state, and local government agencies and related groups.

B. EOP Maintenance

DEM is responsible for maintaining the EOP. The EOP is continually reviewed and periodically updated as required to incorporate federal policy changes, gubernatorial directives, legislative changes, and operational changes based on lessons learned from exercises and actual events. The EOP will be reviewed and adopted in its entirety by the Governor at least every four years. This section outlines protocols for interim changes and full updates of the EOP. An EOP Management Standard Operating Procedure has been adopted to further define these protocols. Changes include additions of new or supplementary material and deletions. No proposed change should contradict or override authorities or other plans.

Any agency or coordinating NGO or PSP may propose and develop a change to the EOP. DEM is responsible for coordinating review of the proposed change among the lead and support agencies, NGOs, and PSPs of each affected ESF and any associated agency program areas as required. If DEM identifies planning needs that require immediate resolution, or at the request of a state agency, DEM may convene an EOP Plan Committee and revise areas of the plan identified by the committee. Interim changes to the EOP that are administrative in nature may be approved by the State Coordinator, however major revisions or full updates will be sent to the Secretary of Public Safety and Homeland Security for review and concurrence prior to submitting the plan to the Governor for approval.

After receiving approval by the Governor, full updates will be promulgated by Executive Order.

The DEM procedure for changes includes:

- Obtaining the official approval for the change from the appropriate officials of the affected agencies, NGOs, and PSPs.
- A process to notify and receive approval from the Governor or his/her designee for all requested changes.
- Ensure appropriate notification is made about the changes and maintain a record of changes.

C. Standards for other Emergency Plans

The EOP, including all annexes, is the core plan for emergency operations, and provides the structures and processes for coordinating incident management activities for human-caused disasters, natural disasters, and other emergencies or events. Following the guidance provided by the NRF, NIMS, and other related documents, the EOP incorporates and/or provides an umbrella configuration for state emergency and incident management plans (with appropriate modifications and revisions) as integrated components of the EOP, as supplements, or as supporting operational plans.

Accordingly, state agencies must incorporate key EOP concepts and operational elements when developing or updating agency specific incident management and response plans. All additional response and recovery plans or procedures developed by agencies should be compatible with the EOP. Agencies are responsible for providing DEM with current agency plans or procedures that support their role in the EOP.

D. Plan Testing, Training, and Exercises

Responsible state agencies, and coordinating NGOs and PSPs, will conduct training to ensure the EOP can be effectively implemented in a timely manner.

DEM, in coordination with the ERT, will conduct an annual exercise or a series of exercises of the EOP. As required by the state Code, one exercise must address, among other issues, a prolonged and widespread loss of electric power.

Exercises may include multiple agencies, jurisdictions, NGOs, and PSPs. In addition, the ERT will participate with FEMA in an annual exercise of the Radiological Response Annex. Any deficiencies, findings, areas recommended for corrective action, or improvement arising from these exercises, or any other exercises coordinated from the EOC, will be considered and corrected by appropriate training, plan update, and/or demonstration in any subsequent exercise or postulated event. DEM has developed and instituted an after action review (AAR) process in which all ERT agencies participate. Local government, NGO, and PSP representatives are encouraged to participate in the AAR process.

E. Plan Implementation

This plan is effective for execution upon and pursuant to the Executive Order promulgating the same. The State Coordinator will ensure that this document is subject to a minimum of an annual maintenance, review, and update based on selective evaluations, AARs, and new guidance.

VII. AUTHORITIES AND REFERENCES

Note that authorities and references for the entire plan are located in Appendix C

- Authorities
- Emergency Services and Disaster Laws References
- The Stafford Act
- National Incident Management System
- National Response Framework
- National Disaster Recovery Framework
- EMAP Standards (2013): 3.1.2, 4.6.1, 4.6.2, 4.6.3, 4.10.1, 4.10.5, 4.14.3

18 Indian Ocean Tsunami 2004 Case Study

On December 26, 2004, there was a massive and sudden movement of the Earth's crust under the Indian Ocean. This earthquake was recorded at magnitude 9 on the Richter scale and, as it happened under the ocean, caused a devastating sea wave called a tsunami.

The epicenter of the earthquake occurred 200 km west of the island of Sumatra in the Indian Ocean. The earthquake itself happened due to the subduction of the Indo-Australian plate under the Eurasian plate. As the Indian plate (part of the Indo-Australian plate) moved underneath the Burma plate (part of the Eurasian plate), the crustal rocks stuck as they moved past one another. At 8 a.m. local time, the pressure build up was too great and the crustal rocks snapped, causing an earthquake. When this happened, the sea floor was pushed upward, displacing a huge volume of water and creating the devastating Tsunami waves.

INDIAN OCEAN TSUNAMI 2004: IMPACT ON LANDSCAPE AND POPULATION

LANDSCAPE

- Some smaller islands in the Indian Ocean were completely destroyed.
- Coastal buildings were completely destroyed making people homeless.
- Fishing villages were completely destroyed.
- Lines of communication were cut off, including phone and electricity power lines.
- Roads and railways were destroyed.
- Fires broke out due to severed water pipes.

PEOPLE

Approximately 250,000 people are estimated to have been killed including many tourists on the beaches of Thailand.

- There was an outbreak of disease such as Cholera due to a lack of fresh water supplies.
- There was a lack of food as many fish died and farms were destroyed.

INDIAN OCEAN TSUNAMI 2004: METHODS OF PREDICTION

Earthquakes are extremely difficult to predict, although scientists now know which areas have a higher risk of earthquakes and can identify frequency patterns from previous earthquakes. As a result, active earthquake zones are closely monitored for seismic activity, including the use of tilt meters and laser equipment to measure earth movements and sophisticated sound recording equipment to monitor earth tremors.

In developed countries such as the United States, building earthquake-proof buildings, such as the Transamerica Pyramid in San Francisco, which sways with the movement of the earth, has helped to reduce the damage caused by earthquakes. Some countries practice earthquake drills, e.g., Tokyo in Japan, just as we have fire drills and emergency services are better prepared and equipped to deal with such a disaster.

Despite all of these measures, there were few warnings or successful predictions of the Indian Ocean Tsunami. Most of the countries affected were developing countries without the funds for these sophisticated methods of detection.

They also lacked the improved communications, which may have allowed them to evacuate from coastal areas in time. All the warning they received was the retreat of sea water from beaches before the wave hit.

INDIAN OCEAN TSUNAMI 2004: ROLE OF AID AGENCIES

Short-Term Aid

- Emergency medical aid was provided by The Red Cross.
- Tents and blankets were provided for the homeless by United Nations (UN) Children's Fund.
- Temporary shelters were constructed.
- Bottled water was supplied as there was restricted access to fresh, clean water.
- Bulldozers, sniffer dogs, and heat seeking equipment were used to detect people who may have been lost or missing in collapsed buildings.
- Food supplies given by the UN due to destruction of farmland and buildings.
- Clothing supplies would be needed.
- Airlifting of more serious cases to hospitals in unaffected areas was done by the U.S. military.

Long-Term Aid

- Roads, electricity supplies, water pipes, etc., needed to be rebuilt.
- Houses and other buildings such as schools and hospitals needed rebuilding.
- Flood prevention measures such as sea walls to be built.

Effectiveness

- There was still a huge loss of life.
- The tourist industry, which was a large source of income, particularly to Thailand and Sri Lanka, was up and running again after only a few months.
- Warning through media saved lives in areas such as East Africa.

- Money was raised initially through TV campaigns, but this dried up once the tsunami was no longer in the news.
- Potential corruption and stories of aid not reaching victims of the disaster

IMPACT OF THE 2004 TSUNAMI IN THE ISLANDS OF INDIAN OCEAN: LESSONS LEARNED

The tsunami of 2004, caused by a 9.0-magnitude earthquake, is the most devastating tsunami in modern times, affecting 18 countries in Southeast Asia and Southern Africa, killing more than 250,000 people in a single day, and leaving more than 1.7 million homeless. However, less reported, albeit real, is its impact in the islands of the Indian Ocean more than 1,000 miles away from its epicenter.

The 2004 tsunami events, specifically in the 11 nations bordering the Indian Ocean, as they constitute a region at risk, due to the presence of tectonic interactive plate, absence of a tsunami warning system in the Indian Ocean, and lack of established communication network providing timely information to that region. Our paper has a dual objective: The first objective is to report the 2004 tsunami event in relation to the 11 nations bordering the Indian Ocean.

The second one is to elaborate on lessons learned from it from national, regional, and international disaster management programs to prevent such devastating consequences of tsunami from occurring again in the future.

INTRODUCTION

Tsunami is a series of ocean waves typically caused by large undersea earthquakes or volcano eruptions at tectonic plate boundaries. These surges of water may reach 100 feet and cause widespread destruction when they crash ashore. They race across the sea at a speed up to 500 miles per hour and cross the entire Pacific Ocean in less than one day. Their long wavelength means that they lose very little energy along the way.

The tsunami of December 2004, caused by a 9.0-magnitude earthquake, is the most infamous tsunami of modern times with disastrous consequences in many areas:

1. (i) Humanitarian toll: It affected more than 18 countries from Southeast Asia to Southern Africa, killing more than 250,000 people in a single day and leaving more than one million homeless. (ii) Economic toll: It left several millions of dollars of economic loss affecting fishing and tourist industries. (iii) Environmental and medical threats including water pollution and flooding and endemic diseases.

   The rationale for writing this paper is to report the tsunami events in the 11 nations bordering the Indian Ocean, as they received less publicity than their Southeast Asian country counterparts, although the 2004 tsunami had real humanitarian, economic, and environmental impact in these regions more than 1,000 miles away from the epicenter.

2. Furthermore, these regions are at risk from the devastating effects of future tsunami due to the presence of a tectonic interactive plate.
3. The absence of a tsunami warning system in the Indian Ocean and lack of established communication network providing timely information to that region.

IMPACT OF THE TSUNAMI IN THE ISLANDS OF THE INDIAN OCEAN

These 11 countries bordering the Indian Ocean are Mauritius, Madagascar, Reunion Island, and Seychelles, Comoros islands, and by geographical extension include countries in southern borders of Africa such as Somalia, Tanzania, Mozambique, and South Africa.

These individual countries suffered humanitarian loss, with more than 3,000 people killed and more than 10,000 left homeless about 1,000 miles away from the epicenter. In terms of economic toll, several million dollars were reported accompanied by environmental threat due to flooding ("Impact of 2004 Tsunami in the Islands of Indian Ocean," https://www.hindawi.com/journals/emi/2011/920813).

LESSONS LEARNED FROM THE 2004 TSUNAMI

To prevent the devastating effects of future tsunami, these islands of Indian Ocean have set their priorities in achieving three goals:

1. Development of a disaster tsunami program that includes implementation of a tsunami program at the national, regional, and international levels and coordination of all these programs.
2. Development of an Indian Ocean early warning system.
3. Development of a tsunami research program.

National Level

This national response plan was developed by Indian in 2006 and is the most exhaustive of all the other national plans and should serve as a model for other islands (National Response Plan, https://fas.org/irp/agency/dhs/nrp.pdf). It includes five objectives:

1. Development of a national evacuation plan on tsunami.
2. Establishment of an early warning system in conjunction with a regional system.
3. Increase public and community awareness through publication and training of media and local authorities.
4. Conduct mock exercises on tsunami.
5. Strengthen the operational capacity of the national meteorological service to include a national warning system.

Regional Level

Disaster management is a regional priority in the Indian Ocean due to permanent threat of cyclones, floods, and tsunamis. The stated two goals set by a series of regional meetings in 2005 and 2006 are the following:

1. Implementation of the Indian Ocean tsunami warning and mitigation system, which focuses on defining disaster management and reduction (prevention, mitigation, response, and relief) of disaster by all the participating countries.
2. Development of the integrated regional information network, with the goals of creation of an early warning system for the islands in the Indian Ocean and ensuring adequate equipment to manage natural disasters, including tsunamis.

The important issues are the cost of establishing such a warning system in the Indian Ocean, the transfer of existing technology versus improving an old one, global warming, and extreme weather events in that region.

International Level

A series of international meetings have been convened to discuss the role international organizations (United Nations Educational, Scientific and Cultural, https://sustainabledevelopment.un.org/content/documents/index.php?):

1. International coordination meeting sponsored by the UN Educational Scientific and Cultural Organization intergovernmental oceanographic commission in Mauritius in 2005 with a dual goal: (a) development of a tsunami warning and mitigation system and (b) coordination of a national tsunami warning center with regional centers.
2. International strategy for disaster reduction (UN/ISDR) attended by representatives from the Indian Ocean countries and international experts on early warning system in 2006 with two objectives:
   (a) Funding of projects and rehabilitation of roads and bridges
   (b) Increase public awareness and training of key staff in tsunami preparedness and warning at all levels.

CONCLUSION

The 2004 tsunami on the islands bordering the Indian Ocean and the lessons learned from this event from national, regional, and international organizations to prevent such events from occurring again in the future are discussed.

A tsunami is an ever-present and real threat for the islands of the Indian Ocean due to the presence of a tectonic interactive plate.

Their disaster management priority is the development of an early tsunami warning system in order to effectively and timely communicate with all the people in that region.

Disaster management should involve national, regional, and international organizations at all levels in order to develop tsunami program, fund tsunami projects, and continue research program.

19 Gulf of Mexico Oil Spill and BP Case Study

On April 20, 2010, a deep-water oil well exploded in the Gulf of Mexico. The immediate effect was that it killed 11 people and injured 17 others. Oil leaked at a high rate, which is difficult to calculate. Some estimates are around 40,000 barrels a day. The oil spill posed risks to the environment and affected the local industry.

The impact of this oil spill depended on which parts of the coastline you look at. It is difficult to measure the effects because of seasonal changes in wildlife.

ECONOMIC IMPACT

- The government asked for $20 billion in damages from British Petroleum (BP) and BP's share price fell. Local industries, such as fishing, were threatened. There was a ban on fishing in the water.
- Tourism declined.

ENVIRONMENTAL IMPACT

Gulf of Mexico oil spill rescue, 2010

- An environmental worker rescued an oil-covered pelican.
- Plants and animals were completely covered in the oil.
- Seabirds, sea turtles, and dolphins were found dead.
- Oil that entered wetland areas meant recovery was slow.
- Fish stocks were harmed, and productivity decreased.

The size of the oil spill was one of the largest America had seen. However, because the oil entered warm waters, organisms in the water helped to breakdown the oil. The overall effect may be less than the Exxon Valdez Oil spill in 1989, which happened in colder water (Figure 19.1).

The cause and effects of the oil spill were immense, and the massive damage to the wetlands expanded the dead zones, killing off marine mammals and blobs of oil coating deep-sea coral and causing illness in dolphins. Immersed underwater plumes of dissolved oil tar balls, also found in fishing nets.

Thc British multinational oil and gas company BP headquartered in London in the United Kingdom owned Deepwater Horizon, which experienced the oil spill in the Gulf of Mexico near Mississippi River Delta in the United States. At approximately 9:45 p.m. on April 20, 2010, an explosion occurred, causing a massive oil spill in the Gulf.

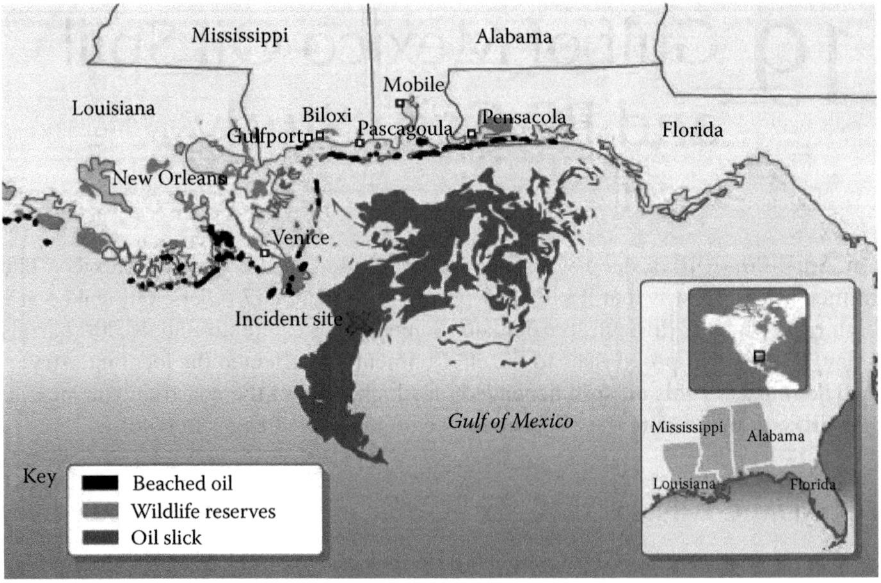

FIGURE 19.1 Map of the oil spill.

The spill caused 4,200 square miles (11,000 km²) of the Gulf to close due to the massive oil spill and the time it took to clean it up.

In the efforts to clean the spill up and monitor its effects:

- BP made use of skimmers to clean up the oil spill. They also tested the new gigantic oil skimmer called "the Whale," which can clean up a large quantity of water.
- They gave money to people who lost their livelihood, as reparation for the spill, and they looked forward to finding solutions to avoid similar spills.
- BP was still paying for commercials encouraging toursim to the Gulf of Mexico as late as 2012, two years after the disaster.
- BP aims to maintain readiness to respond on a global scale, to minimize adverse effects and facilitate rapid mitigation activities. Some companies from the United States helped in the cleanup process of some affected animals, e.g., Dawn (dishwashing soap company). Local birds and animals were muddied in toxic sludge. Volunteers and rescue used a common household ingredient to clean up the oil-coated wildlife—Dawn dishwashing liquid.
- People across the United States were ready to dedicate their time and ideas to clean up the oil at the coasts. They cleaned up the Gulf Coast, and they had crews cleaning up the beaches, as well as skilled workers taking care of animals.

- The U.S. Coast Guard also set up a base in St. Petersburg, Florida, to track and coordinate cleanup efforts for any spilled oil that touches the state's west coast.

LESSONS LEARNED FROM THE BP OIL SPILL

The fallout from this disaster has raised many questions about both the history and future of oil drilling and our future in energy resources as a country.

Releasing an estimated 4.4 million barrels, or 185 million gallons, of oil into the Gulf of Mexico, BP's oil spill is now a tragic example of the need for better energy technologies. To understand the current situation, researchers have been investigating past spills and drilling history, as ways to estimate the potential impacts the Gulf will still have in the years ahead.

Seven distinct areas of critical issues are as follows:

- Improving the safety of offshore operations
- Safeguarding the environment
- Strengthening oil spill response, planning, and capacity
- Advancing well-containment capabilities
- Overcoming the impacts of the deepwater horizon spill and restoring the gulf
- Ensuring financial responsibility of responsible parties
- Promoting congressional engagement to ensure responsible offshore drilling

Three critical issues as having relevance to efforts to strengthen its oil spill prevention, preparedness, and response system are the following:

- Improving the safety of offshore operations
- Strengthening oil spill response, planning, and capacity
- Ensuring financial responsibility

Like many states around the nation, the question is often asked here—how prepared is Washington for a catastrophic spill? The reality here and around the nation is that no matter how well prepared we are for spills, we fight a losing battle from the start and we have to rely on other regions to assist in our response and recovery to a Spill of National Significance.

Recognizing that preparedness is a continuous cycle, we must put in place now the mechanism and processes to the following:

- Address the risk of oil spills and shift priorities as that risk evolves.
- Analyze the adequacy of and access to appropriate response equipment.
- Continuously find and fix the weak areas of the system through testing of plans.

- Promote and encourage response technology development.
- Strengthen and enhance relationships with our federal, state, local, tribal industry, and community partners.

IMPROVING THE SAFETY OF OFFSHORE OPERATIONS

The Need for a New Approach to Risk Assessment and Management

> Neither the industry's nor the federal government's approaches to managing and overseeing the leasing and development of offshore resources have kept pace with rapid changes in the technology, practices, and risks associated with the different geological and ocean environments being explored and developed for oil and gas production ("Improving Oil Spill Prevention and Response in Washington," http://www.ecy.wa.gov/programs/spills/studies_reports/ecypspreview-dwh).

There are operational risks inherent in oil transportation and refining activities taking place on our state waters and lands. Preventing large and small spills is Washington's top priority and a legislative goal ("zero spills"). Meeting this goal requires the EPA agency to identify industry-specific risks at every level in marine transportation and oil handling systems and to target those risks with the right prevention activities.

Key prevention issues, such as fatigue, inadequate crewing requirements and inadequate company training, operating procedures, and policies continue to be prominent causal factors in these incidents. Ecologists must continue working with industry partners to emphasize human factors as the key to spill prevention. The data also indicate that incidents from nonregulated sources, such as fishing and recreational vessels, pose a disproportionate risk of incidents and spills.

Recommendations

- Complete a rigorous risk analysis on higher-risk industry sectors to ensure that there is an appropriate level of investment reducing the risk of oil spills. Target our spill prevention education, grant assistance, inspection, and regulatory activities based on the risks presented by the various sectors.
- Complete an analysis of U.S. Coast Guard (USCG) and Ecology tug/oil barge incident data. This industry sector appears to have a relatively high risk of towing incidents in which the oil barges contain millions of gallons.
- Expand vessel inspection activities of regulated fishing vessels and ensure that nonregulated fishing vessels receive voluntary pollution control technical assistance.
- Continue to seek opportunities to influence federal standards relating to these risk factors. Seek formal or informal delegation from the USCG of some activities where the agencies have concurrent jurisdiction and program missions.

STRENGTHENING OIL SPILL RESPONSE, PLANNING, AND CAPACITY

The Area Plan contains response policies and tools and provides a coordination mechanism for all three states and the federal government. The Area Plan is updated annually. Industry oil spill plans are reviewed for their consistency with the Area Plan and are measured against regulatory standards for approval.

There are many complex coordination and logistical issues that must be addressed well ahead of a spill. The goal must be to mount a response that is rapid, aggressive, and well coordinated with our Canadian partners in industry, the federal government, and province.

Recommendations

- Once the Pacific States/British Columbia Oil Spill Task Force report is issued, response agencies in the Trans-boundary area should develop and implement an action plan around the findings in the Tran's boundary Spill Planning and Response Issues report.
- The Joint Response Teams should work quickly to resolve gaps in coordination mechanisms and adopt consistent policies and processes to manage cross-border spills.

THE NEED FOR A NEW APPROACH TO HANDLING SPILLS OF NATIONAL SIGNIFICANCE

Recommendations

- Ecology will adopt the methodology and process for making decisions on equipment movement during spills of national significance (SONS) into the Area Plan as a best management practice.
- The federal government should develop an alternative process to ensure rapid cascading of equipment other than the emergency rule standard used in the Gulf SONS.
- The federal government should develop a SONS process mechanism to engage in Tribal Consultation when tribal resources are impacted by a SONS designation.

THE NEED TO STRENGTHEN STATE AND LOCAL INVOLVEMENT

Local elected official, tribal government, and citizen involvement in the oil spill planning process is critical if an area is to be truly prepared to deliver a well-coordinated response.

This involvement ensures that the response team is able to take advantage of local knowledge and local resources in the response. The Area Plan has a longstanding policy to include tribes and local government in Unified Command. Beyond decision-making processes, local involvement also includes volunteer management and assisting in the identification of local commercial vessels of opportunity for use as response assets.

Local Involvement in Oil Spill Drills

Drills are a requirement for companies to test their plans. They are also an opportunity to involve local officials and tribes in the Area Plan, so that the coordination and prioritization issues experienced in the Gulf do not occur here. Area Plan initiatives are also an opportunity for companies and response agencies to learn local knowledge that is critical during a spill response. All parties to a spill need to know other people involved in the spill response prior to the day of the spills. This requires practice. We recognize that funding limitations for tribes and local government is a limiting factor to their involvement.

Volunteer Management Program

Many Washington State citizens feel outraged and frustrated when oil spills impact their beaches, and they feel compelled to take action. For many, this means looking for opportunities to be involved as volunteers and contribute to restoring their community. Through the Beach Watchers program, Ecologists has provided training to community groups around the state who serve as "eyes" for Ecology and in many cases are the closest field observers to help size up reported spills (Beach Watchers | Snohomish MRC, www.snocomrc.org/Projects/Education-Outreach/Beach-Watchers.aspx).

Many volunteers are interested in participating with oiled wildlife care. Because of the potential to harm affected wildlife and the potential for human exposures to toxic oil and diseases, additional levels of training are required. This training is coordinated by the Washington Department of Fish and Wildlife and federal agencies within an existing network of wildlife care organizations.

There are a number of issues to be addressed regarding use of volunteers during oil spill response, including liability, proper training, compensation, reporting, and supervision relationships. Similar to the need to have local governmental and tribal involvement prior to a large spill, it is also critical to have trained and organized volunteers to be accessible during spill incidents. While managing volunteers is an inherently governmental function, funding for use of volunteers should be borne by the industry during an actual spill incident.

Vessel of Opportunity Program

In a major spill, it is evident early on that the need for vessels to support response activities over large geographic areas will outgrow the professional, dedicated vessel response assets. The Vessel of Opportunity Program (VOO) is an opportunity to utilize existing vessels, such as fishing and other commercial vessels in our waters during oil spill response (Factsheet on BP Vessels of Opportunity Program, https://publicintelligence.net/wp-content/uploads/group-documents). The Gulf oil spill offered many lessons learned in using local fishing fleets and other commercial vessels as spill response assets.

The VOO Program offers assistance in many response activities, including the following:

- Transporting supplies and providing overnight berthing (lodging).
- Assisting wildlife survey and rescue.
- Deploying containment and sorbent boom.
- Providing on-water recovery (skimming) and storage.

There are many viable advantages to using vessels of opportunity, such as accessing local knowledge of waterways and environmental conditions. It also offers employment opportunities for an economic sector that may be severely impacted by oil spills.

The reality of the Gulf oil spill is that as long as we are dependent on oil as our main energy resource, we are required and obligated to ensure that oil exploration and refining and transport of petroleum are safe and our environment and economy are protected from oil spills.

The Gulf area is heavily dependent on maritime commerce, the marine environment, and the continued recovery and restoration of the Gulf are key elements of its economic engine. It is recognized that implementing some of these recommendations may be challenging in the state's current economic environment.

However, it is critical that our state make appropriate investments now to ensure that we protect the state's future economic interests against major oil spills.

The recommendations outlined in this report promise effective protections to avert or mitigate damages from a major oil spill and avoid the potential of an environmental disaster that would threaten our economy, our environment, and our region's quality of life.

Appendix A: Terms and Definitions

Terms and definitions in the Federal Response Plan generally are consistent with current terminology used in the emergency management community. A number of these terms are defined below. Others are defined in the Basic Plan and individual annexes. Many of these terms are cross-referenced in this appendix.

LIST OF TERMS

Accountable Property: Action Plan. See ESF 5.
Aerial Port of Debarkation: See ESF 9.
Aerial Port of Embarkation: See ESF 9.
After-Action Report
Asset Visibility
Assets
Base Support Installation: See ESF 9.
Biological Agents
Catastrophic Disaster Response Group (CDRG)
Chemical Agents
Civil Air Patrol (CAP): See ESF 5.
Civil Transportation Capacity: See ESF 1.
Comprehensive Environmental Response, Compensation, and Liability Act (CERCLA): See ESF 10.
Congressional Affairs Representative (CAR)
Congressional Relations Officer (CRO)
Consequence Management
Contingency Plan: See ESF 5.
Credible Threat
Crisis Management
Defense Coordinating Officer (DCO)
Designated Agency Safety and Health Official (DASHO)
Disaster Field Office (DFO)
Disaster Finance Center (DFC)
Disaster Information Systems Clearinghouse (DISC)
Disaster Medical Assistance Team (DMAT): See ESFs 8 and 9.
Disaster Mortuary Team (DMORT): See ESF 8.
Disaster Recovery Center (DRC)
Disaster Recovery Manager (DRM)
Disaster Response Support Facility (DRSF)
Disaster Safety Officer (DSO)

Disaster Transportation Management System (DTMS): See ESF 1.
District Response Group: See ESF 10.
Domestic Emergency Support Team (DEST)
Donations Coordination Center
Donations Coordination Team
DOT Crisis Coordinator: See ESF 1.
Emergency Response Team (ERT)
Emergency Response Team—Advance Element (ERT-A)
Emergency Support Function (ESF)
Emergency Support Function Leaders Group (ESFLG)
Emergency Support Team (EST)
Environmental Response Team: See ESF 10.
Essential Elements of Information (EEIs): See ESF 5.
Federal Coordinating Officer (FCO)
Federal Emergency Support Coordinator (FESC): See ESF 7.
Federally Arranged Transportation Support: See ESF 1.
FEMA Voluntary Agency Liaison (VAL)
Fire Suppression Support Coordinator: See ESF 4.
Food and Nutrition Service (FNS) Disaster Task Force: See ESF 11.
Functional Plan: See ESF 5.
Goods
Governor's Authorized Representative (GAR)
Hazardous Materials: See ESF 10.
Hazardous Substances: See ESF 10.
Incident Command System (ICS): ESF 4.
Incident Support Team (IST): See ESF 9.
Incident Support Team—Advance Element (IST-A): See ESF 9.
Information Coordination Unit (ICU): See ESF 5.
Initial Response Resources (IRR)
In-Kind Donations: See Donations Management Support Annex.
Joint Information Center (JIC)
Joint Operations Center (JOC)
Lead Agency
Lead Federal Agency
Logistics Information Management System (LIMS): See ESF 5.
Memorandum of Agreement (MOA): See ESF 9.
Mobilization Center: See ESF 9.
Monitoring Period: See ESF 5.
Movement Coordination Center (MCC): See ESF 1.
National Disaster Medical System (NDMS): See ESF 8.
National Fire Suppression Liaison Officer: See ESF 4.
National Interagency Coordination Center (NICC): See ESF 4.
National Oil and Hazardous Substances Pollution Contingency Plan (NCP): See ESF 10.
National Processing Service Center (NPSC)

National Response Center (NRC): See ESF 10.
National Response Team (NRT): See ESF 10.
National Security Council (NSC)
National Strike Force: See ESF 10.
National Voluntary Organizations Active in Disaster (NVOAD)
Nuclear Weapons
On-Scene Coordinator (OSC): See ESF 10.
Operating Site: See ESF 9.
Operational Period: See ESF 5.
Personal Property
Preliminary Damage Assessment (PDA)
Presidential Decision Directive 39 (PDD-39): Primary Agency. See Basic Plan, pages 13 and 28.
Radiological Emergency Response Team: See ESF 10.
Reconstruction Information Center (RIC)
Regional Emergency Coordinator (REC): See ESF 7.
Regional Emergency Transportation Coordinator (RETCO): See ESF 1.
Regional Operations Center (ROC): See Basic Plan, page 17.
Regional Response Teams (RRTs): See ESF 10.
Regional/Area Fire Coordinator: See ESF 4.
Requirements Processing
Resource Tracking
Resources
Scientific Support Coordinator: See ESF 10.
Senior FEMA Official
Situation Assessment: See ESF 5.
Situation Report (SITREP): See ESF 5.
Situation Room: See ESF 5.
Staging Area: See ESF 9.
State Coordinating Officer (SCO)
Status Briefing: See ESF 5.
Strategic Information and Operations Center (SIOC)
Strategic Plan: See ESF 5.
Supervisor of Salvage and Diving (SUPSALV): See ESF 10.
Support Agency
System to Locate Survivors (STOLS): See ESF 9.
Technical Operations
Territory Logistics Centers (TLCs)
Terrorist Incident: Time-Phased Force and Deployment List (TPFDL). See ESF 1.
Unaffiliated Volunteer
Undesignated Goods
Unsolicited Goods
Voluntary Organizations Active in Disaster (VOAD)
Weapon of Mass Destruction (WMD)

DEFINITIONS

Agency Logistics Center (ALC): See Logistics Management Support Annex.

Assembly Point: A designated location for responders to meet, organize, and prepare their equipment prior to moving to the point of departure. Since emergency teams, organizations, and resources involved in a disaster or emergency can originate from a variety of geographic locations, each typically has its own assembly point.

Base Camp: The designated location under local or state control within the disaster area that is equipped and staffed to provide sleeping facilities, food, water, and sanitary services to response personnel.

Designated Area: The geographic area designated under a Presidential major disaster declaration that is eligible to receive disaster assistance in accordance with the provisions of the Stafford Act.

Direct Federal Assistance: Is provided to the affected state and local jurisdictions when they lack the resources to provide specific types of disaster assistance either because of the specialized nature of the assistance or because of resource shortfalls (e.g., providing debris removal, potable water, emergency medical services, and urban search and rescue).

Emergency: As defined in the Stafford Act, an emergency is any occasion or instance for which, in the determination of the president, Federal assistance is needed to supplement state and local efforts and capabilities to save lives and to protect property, public health, and safety and includes emergencies other than natural disasters.

Federal Operations Support: Is available to FEMA or other Federal responding agencies when they require logistical or technical support of their Federal operations—ESF activation, personnel for preparing damage survey reports, supplies, and equipment for DFO and DRC operations.

Major Disaster: As defined under the Stafford Act, any natural catastrophe (including any hurricane, tornado, storm, high water, wind-driven water, tidal wave, tsunami, earthquake, volcanic eruption, landslide, mudslide, snowstorm, or drought), or, regardless of cause, any fire, flood, or explosion, in any part of the United States, which in the determination of the president causes damage of sufficient severity and magnitude to warrant major disaster assistance under this act to supplement the efforts and available resources of states, local governments, and disaster relief organizations in alleviating the damage, loss, hardship, or suffering caused thereby.

Mitigation: Those activities designed to alleviate the effects of a major disaster or emergency or long-term activities to minimize the potentially adverse effects of future disaster in affected areas.

Point of Arrival (POA): The designated location (typically an airport) within or near the disaster-affected area where newly arriving staff, supplies, and equipment are initially directed. Upon arrival, personnel and other resources are dispatched to either the DFO, a mobilization center, a staging area, or directly to a disaster site.

Point of Departure (POD): The designated location (typically an airport) outside the disaster-affected area from which response personnel and resources will deploy to the disaster area.

Recovery: Activities traditionally associated with providing Federal supplemental disaster relief assistance under a presidential major disaster declaration. These activities usually begin within days after the event and continue after response activity ceases. Recovery includes individual and public assistance programs that provide temporary housing assistance, as well as grants and loans to eligible individuals and government entities to recover from the effects of a disaster.

Response: Activities to address the immediate and short-term effects of an emergency or disaster. Response includes immediate actions to save lives, protect property, and meet basic human needs. Based on the requirements of the situation, response assistance will be provided to an affected state under the FRP using a partial activation of selected ESFs or the full activation of all ESFs to meet the needs of the situation.

Technical Assistance: Is provided to state and local jurisdictions when they have the resources but lack the knowledge and skills needed to perform a required activity (such as mobile-home park design and hazardous material assessments).

Appendix B: Acronyms and Abbreviations

AE	Aeromedical evacuation
AEC	Agency emergency coordinator
AECC	Aeromedical evacuation control center
AECE	Aeromedical evacuation control element
AECM	Aeromedical evacuation crew member
AELT	Aeromedical evacuation liaison team
AID	Agency for International Development
ALC	Agency logistics center
AMC	Air mobility command
AMTA	Agricultural Marketing Transition Act
AOC	Army operations center
ARAC	Atmospheric Release Advisory Capability
ARC	American Red Cross
ASH	Assistant secretary for health
AWD	Available without declaration
B&I	Business and Industrial Loan Program
C/B	Chemical/biological
CAP	Civil Air Patrol
CAR	Congressional affairs representative
CC	Coordination center
CCP	Casualty collection point
CCP	Crisis Counseling Assistance and Training Program
CDBG	Community Development Block Grant
CDC	Centers for Disease Control and Prevention
CDRG	Catastrophic Disaster Response Group
CEPPO	Chemical Emergency Preparedness and Prevention Office
CERCLA	Comprehensive Environmental Response, Compensation, and Liability Act
CFO	Chief financial officer
CFR	Code of Federal Regulations
CINCLANT	Commander-in-Chief Atlantic
CINCPAC	Commander-in-Chief Pacific
CLO	Congressional liaison officer
CMC	Crisis management center
CNS	Corporation for national service
CONUS	Continental United States
CPD	Community planning and development
CR	Community relations

CRO	Congressional relations officer
CRP	Conservation Reserve Program
CWA	Clean Water Act
DAE	Disaster assistance employee
DALO	Disaster area liaison officer
DASHO	Designated agency safety and health official
DCE	Defense coordinating element
DCLO	Deputy congressional liaison officer
DCO	Defense coordinating officer
DEST	Domestic emergency support team
DFC	Disaster finance center
DFCO-M	Deputy federal coordinating officer for mitigation
DFO	Disaster field office
DISC	Disaster information systems clearinghouse
DLA	Defense logistics agency
DMAT	Disaster medical assistance team
DMORT	Disaster mortuary team
DOC	Department of Commerce
DOD	Department of Defense
DOE	Department of Energy
DOEd	Department of Education
DOI	Department of the Interior
DOJ	Department of Justice
DOL	Department of Labor
DOMS	Director of Military Support
DOS	Department of State
DOS-A/DCP	Department of State, Office of Diplomatic Contingency Programs
DOT	Department of Transportation
DRC	Disaster recovery center
DRF	Disaster relief fund
DRM	Disaster recovery manager
DRSF	Disaster response support facilities
DSO	Disaster safety officer
DTMS	Disaster transportation management system
DUA	Disaster unemployment assistance
DWI	Disaster welfare information
EC	Emergency coordinator
ECS	Emergency communications staff
ECWAG	Emergency community water assistance grants
EDA	Economic Development Administration
EEI	Essential element of information
EICC	Emergency Information and Coordination Center
EIDL	Economic injury disaster loans
EM	Emergency management
EMRT	Emergency medical response team

EMS	Emergency medical service
EMT	Emergency management team
EMWIN	Emergency Managers' Weather Information Network
E.O.	Executive order
EOC	Emergency operations center
EPA	Environmental Protection Agency
EPLO	Emergency preparedness liaison officer
ERL	Environmental Research Laboratories
ERT	Emergency response team
ERT	Environmental response team
ERT-A	Emergency response team—advance element
ERT-N	National emergency response team
ESF	Emergency support function
ESFLG	Emergency Support Function Leaders Group
ESP	Electric service priority
EST	Emergency support team
EWP	Emergency watershed protection
FAO	Federal approving official
FAR	Federal acquisition regulation
FBI	Federal Bureau of Investigation
FCC	Federal Communications Commission
FCC	Federal coordinating center
FCO	Federal coordinating officer
FDIC	Federal Deposit Insurance Corporation
FECC	Federal emergency communications coordinator
FEMA	Federal Emergency Management Agency
FESC	Federal emergency support coordinator
FHWA	Federal Highway Administration
FNS	Food and nutrition service
FORSCOM	Forces Command
FRERP	Federal radiological emergency response plan
FRMAC	Federal Radiological Monitoring and Assessment Center
FRP	Federal response plan
FS	Forest service
FSA	Farm Service Agency
FTS	Federal telecommunications service
GAO	Government Accounting Office
GAR	Governor's authorized representative
GIS	Geographic information system
GPMRC	Global Patient Movement Requirements Center
GSA	General Services Administration
HHS	Department of Health and Human Services
HMGP	Hazard Mitigation Grant Program
HQ	Headquarters
HQAMC	Headquarters Air Mobility Command
HQUSACE	Headquarters U.S. Army Corps of Engineers

HQUSAF	Headquarters U.S. Air Force
HSO	Human services officer
HUD	Department of Housing and Urban Development
ICS	Incident Command System
ICU	Information Coordination Unit
IFG	Individual and Family Grant
IMA	Individual mobilization augmentee
IRR	Initial response resources
IRS	Internal Revenue Service
IRT	Initial response team
IST	Incident support team
IST-A	Incident support team—advance element
J-4/JCS	Medical Readiness Division, Office of the Joint Chiefs of Staff
JIC	Joint Information Center
JOC	Joint Operations Center
JPMT	Joint Patient Movement Team
JRMPO	Joint Regional Medical Planning Office
JTF	Joint Task Force
JTPA	Job Training Partnership Act
JTRB	Joint Telecommunications Resources Board
LFA	Lead Federal Agency
LIMS	Logistics Information Management System
MA	Mission assignment
MAC	Mapping and Analysis Center
MAC	Mission assignment coordinator
MASF	Mobile Aeromedical Staging Facility
MATTS	Mobile Air Transportable Telecommunications System
MCC	Movement Coordination Center
MERRT	Medical emergency radiological response team
MERS	Mobile emergency response support
MOA	Memorandum of Agreement
MOC	MERS Operations Center
MOU	Memorandum of understanding
MREs	Meals-ready-to-eat
MSCA	Military Support to Civil Authority
MSU	Management support unit
NASA	National Aeronautics and Space Administration
NCC	National Coordinating Center
NCP	National contingency plan
NCS	National communications system
NCS/DISA-GOSC	NCS/Defense Information Systems Agency—Global Operations Security Center
NCSRM	NCS regional manager
NDMS	National disaster medical system
NECC	National Emergency Coordination Center

NEPA	National Environmental Policy Act
NFIP	National Flood Insurance Program
NGB	National Guard Bureau
NICC	National Interagency Coordination Center
NIFC	National Interagency Fire Center
NIMA	National Imagery and Mapping Agency
NIST	National Institute of Standards and Technology
NOAA	National Oceanic and Atmospheric Administration
NOS	National Ocean Service
N/P	Not provided
NPSC	National Processing Service Center
NRC	National Response Center
NRC	Nuclear Regulatory Commission
NRCS	Natural Resources Conservation Service
NRS	National Oil and Hazardous Substances Response System
NRT	National response team
NS/EP	National security/emergency preparedness
NSC	National Security Council
NSF	National Strike Force
NTSP	National telecommunications support plan
NVOAD	National Voluntary Organizations Active in Disaster
NWR	National Oceanic and Atmospheric Administration Weather Radio
NWS	National Weather Service
OC	Operations center
OCHAMPUS	Office of Civilian Health and Medical Program of the Uniformed Services
OCONUS	Outside the continental United States
OEP	Office of Emergency Preparedness
OET	Office of Emergency Transportation
OFM	Office of Financial Management
OIG	Office of Inspector General
OMB	Office of Management and Budget
OPA	Oil Pollution Act
OPAC	Online payment and collection system
OPHS	Office of Public Health and Science
OPM	Office of Personnel Management
OSC	On-scene commander
OSC	On-scene coordinator
OSC	Operations Support Center
OSHA	Occupational Safety and Health Administration
OSTP	Office of Science and Technology Policy
PAO	Public affairs officer
PBS	Public buildings service
PD	Presidential Declaration
PDA	Preliminary damage assessment

PDASH	Principal Deputy Assistant Secretary for Health
PDD	Presidential Decision Directive
PMS	Payments management system
PNP	Private nonprofit
POA	Point of arrival
POC	Point of contact
POD	Point of departure
PPE	Personal protective equipment
PRA	Patient reporting activity
RAP	Radiological Assistance Program
RC&D	Resource conservation and development
RCP	Region Oil and Hazardous Pollution Contingency Plan
RD	Regional director
REAC/TS	Radiation emergency assistance center/training site
REC	Regional emergency coordinator
RECP	Regional emergency services communications planner
REP	Regional evacuation point
REPLO	Regional emergency preparedness liaison officer
RERT	Radiological emergency response team
RETCO	Regional emergency transportation coordinator
RFA	Request for Federal assistance
RHA	Regional health administrator
RHS	Rural housing service
RIC	Reconstruction Information Center
RISC	Regional Interagency Steering Committee
RMA	Risk Management Agency
ROC	Regional Operations Center
RR	Response and recovery
RRIS	Rapid response information system
RRT	Regional response team
RSPA	Research and Special Programs Administration
RTF	Response task force
RUS	Rural utilities service
SA	Supplemental appropriation
SARA	Superfund Amendments and Reauthorization Act
SBA	Small business administration
SCO	State coordinating officer
SIOC	Strategic information and operations center
SITREP	Situation report
SOP	Standard operating procedure
SSA	Social Security Administration
SSC	Scientific support coordinator
STOLS	System to locate survivors
SUPSALV	Supervisor of salvage and diving
TAES	Tactical aeromedical evacuation system

TIMACS	Telecommunications information management and control system
TLC	Territory Logistics Center
TPFDD	Time-phased force and deployment data
TPFDL	Time-phased force and deployment list
TREAS	Department of the Treasury
TSP	Telecommunications Service Priority
TVA	Tennessee Valley Authority
UC	Unified Command
U.S.	United States
U.S.C.	U.S. Code
US&R	Urban Search and Rescue
USACE	U.S. Army Corps of Engineers
USACOM	U.S. Atlantic Command
USCG	U.S. Coast Guard
USDA	U.S. Department of Agriculture
USPACOM	U.S. Pacific Command
USPS	U.S. Postal Service
USSOUTHCOM	U.S. Southern Command
USTRANSCOM	U.S. Transportation Command
VA	Department of Veterans Affairs
VAL	Voluntary agency liaison
VIP	Very important person
VISTA	Volunteers in service to America
VOAD	Voluntary Organizations Active in Disaster
VOLAG	Voluntary agency
WMD	Weapon of mass destruction

Emergency Support Functions and Emergency Management Planning

EMERGENCY SUPPORT FUNCTIONS

Emergency Support Functions (ESFs) is the grouping of governmental and certain private sector capabilities into an organizational structure to provide support, resources, program implementation, and services that are most likely needed to save lives, protect property and the environment, restore essential services and critical infrastructure, and help victims and communities return to normal following domestic incidents.

- ESF1 Transportation
- ESF2 Communications
- ESF3 Public Works and Engineering
- ESF4 Firefighting
- ESF5 Emergency Management
- ESF6 Mass Care, Housing, and Human Services
- ESF7 Resources Support
- ESF8 Public Health and Medical Services
- ESF9 Urban Search and Rescue
- ESF10 Oil and Hazardous Materials Response
- ESF11 Agriculture and Natural Resources
- ESF12 Energy
- ESF13 Public Safety and Security
- ESF14 Long-term Community Recovery and Mitigation
- ESF15 External Affairs

ESF 1—Transportation

- ESF Coordinator: Department of Transportation
- Aviation/airspace management and control
- Transportation safety
- Restoration and recovery of transportation infrastructure
- Movement restrictions
- Damage and impact assessment

ESF 2—Communications

- ESF Coordinator: Department of Homeland Security (DHS) (National Communications System)
- Coordination with telecommunications and information technology industries
- Restoration and repair of telecommunications infrastructure
- Protection, restoration, and sustainment of national cyber and information technology resources
- Oversight of communications within the Federal incident management and response structures

ESF 3—Public Works and Engineering

- ESF Coordinator: Department of Defense (U.S. Army Corps of Engineers)
- Infrastructure protection and emergency repair
- Infrastructure restoration
- Engineering services and construction management
- Emergency contracting support for life-saving and life-sustaining services

ESF 4—Firefighting

- ESF Coordinator: Department of Agriculture (U.S. Forest Service)
- Coordination of Federal firefighting activities
- Support to wildland, rural, and urban firefighting operations

ESF 5—Information and Planning

- ESF Coordinator: DHS (Federal Emergency Management Agency [FEMA])
- Collects, analyzes, processes, and disseminates information about a potential or actual incident
- Conducts planning activities

ESF 6—Mass Care, Emergency Assistance, Temporary Housing, and Human Services

- ESF Coordinator: DHS (FEMA)
- Mass care
- Emergency assistance
- Disaster housing
- Human services

ESF 7—Logistics Management and Resource Support

- ESF Coordinators: General Services Administration and DHS (FEMA)
- Comprehensive, national incident logistics planning, management, and sustainment capability

- Resource support (facility space, office equipment and supplies, contracting services, etc.)

ESF 8—Public Health and Medical Services

- ESF Coordinator: Department of Health and Human Services
- Public health
- Medical
- Mental health services
- Mass fatality management

ESF 9—Search and Rescue

- ESF Coordinator: DHS (FEMA)
- Life-saving assistance
- Search and rescue operations

ESF 10—Oil and Hazardous Materials Response

- ESF Coordinator: Environmental Protection Agency
- Oil and hazardous materials (chemical, biological, radiological, etc.) response
- Environmental short- and long-term cleanup

ESF 11—Agriculture and Natural Resources

- ESF Coordinator: Department of Agriculture
- Nutrition assistance
- Animal and plant disease and pest response
- Food safety and security
- Natural and cultural resources and historic properties protection
- Safety and well-being of household pets

ESF 12—Energy

- ESF Coordinator: Department of Energy
- Energy infrastructure assessment, repair, and restoration
- Energy industry utilities coordination
- Energy forecast

ESF 13—Public Safety and Security

- ESF Coordinator: Department of Justice
- Facility and resource security
- Security planning and technical resource assistance
- Public safety and security support
- Support to access, traffic, and crowd control

ESF 14—Long-Term Community Recovery

- Superseded by the National Disaster Recovery Framework (NDRF)
- Long-Term Community Recovery was superseded by the NDRF. For guidance on long-term community recovery, please refer to the NDRF. http://www.fema.gov/national-disaster-recovery-framework Refer to this link for more information on ESF 14: http://www.fema.gov/media-library/assets/documents/32222?id=7368

ESF 15—External Affairs

- ESF Coordinator: DHS
- Emergency public information and protective action guidance
- Media and community relations
- Congressional and international affairs
- Tribal and insular affairs

EMERGENCY MANAGEMENT PLANNING—FREQUENTLY ASKED QUESTIONS

1. *What is emergency management (EM) planning?*
    - EM planning is a systematic approach for identifying and minimizing the impact of risks to life, property, and the environment.
    - EM planning provides the foundation for coordinating and integrating all activities necessary to build, sustain, and improve the capability to mitigate against, prepare for, respond to, and recover from threatened or actual natural disasters, criminal acts including terrorism, or other manmade disasters.
2. *Why should federal/state/local institutions undertake EM planning?*

   Emergency managers are responsible for developing, testing, and maintaining mandate-specific EM plans and identifying risks that are within or related to their area of responsibility.
    - EM planning promotes a common understanding of the risks, threats, and vulnerabilities that may impact an organization and its related sectors and integrates strategies for addressing possible situations.
    - EM planning will help federal institutions develop more comprehensive and robust systems and practices in EM which will, in turn, help strengthen the Government of Canada system.
    - EM planning provides a systematic way to think through the life cycle of a potential event, determine the required capabilities and resources, and help stakeholders learn and practice their roles.
3. *What is the EM Planning Guide for?*

   The EM Planning Guide (the Guide) is to assist government institutions in the preparation of an EM plan and also in response to the recommendations made by the NIMS/FEMA/NRP and other institutions on EM.

- The Guide furthers the objectives of the EM and, together with the Federal Policy for Emergency Management, serves to assist in strengthening resiliency by promoting a coordinated approach and a more uniform structure across federal/state/local government institutions.
- Its purpose is to assist all federal/state/local institutions in meeting their responsibilities to identify the risks that are within or related to their area of responsibility—including those related to critical infrastructure—and to provide those institutions with a step-by-step approach to develop their all-hazards Strategic Emergency Management Plans (SEMP).

4. *What is the planning premise of EM planning?*
   - EM planning should be based on all four pillars of EM and a comprehensive All-Hazards Risk Assessment.
5. *What are the "4 pillars" of EM?*
   1. *Prevention and mitigation:* Prevention and mitigation refers to actions taken to identify and reduce the impacts and risks of hazards before an emergency or disaster occurs.
   2. *Preparedness:* Preparedness increases the ability to respond quickly and effectively to emergencies and to recover more quickly from their long-term effects and involves actions taken prior to an event to ensure the capability and capacity to respond.
   3. *Response:* Response refers to actions taken during or immediately after an emergency or disaster for the purpose of managing the consequences.
   4. *Recovery:* Recovery refers to actions taken after an emergency or disaster to reestablish or rebuild conditions and services.
6. *When should I begin EM planning?*
   - Optimally, the timing for EM planning would correspond with the environmental-related risk assessment activities.
   - Your SEMP should be linked to your EM planning and reflective of your established priorities.
7. *What is a SEMP?*
   - A SEMP establishes a federal/state/local government institution's objectives, approach, and structure that set out how the institution will assist the coordinated federal emergency response and should, if applicable, include any Emergency Support Functions.
   - EM plans, such as the SEMP, represent an institution's planning associated with its "external" environment.
   - The qualifier "Strategic" is used to differentiate this high-level plan from other types of EM plans, including operational plans. The development and implementation of a SEMP are important complements to other types of EM plans because it promotes an integrated and coordinated approach to EM planning.
8. *Are there other types of EM plans?*

   Yes. There are several different types of EM plans:
   - Operational Plans
   - Response Plans

- Incident-/Risk-Specific Plans (i.e., fire)
- Event-Specific Plans (i.e., G8)
- Institution or Division Specific Plans (i.e., Regional Plan)
- Business Continuity Plans

9. *What is meant by All-Hazards Risk Assessment?*

All-Hazards Risk Assessment is a systematic approach for concurrently identifying, analyzing, and estimating all natural, accidental, and malicious threats and hazards.

All-Hazards Risk Assessment answers four key questions in a systematic way, as a basis for all EM planning:

- What can go wrong and how much warning time are we likely to have?
- How likely is it that the risk will happen?
- What are the consequences for specific stakeholders and society as a whole?
- How well prepared is the institution, collectively, to respond and recover from such a risk?

By addressing these questions systematically, an All-Hazards Risk Assessment

- Promotes the development of a management structure, processes, and procedures throughout the four phases of EM that are applicable to every significant identified hazard;
- Helps to balance and prioritize risk investments and actions;
- Helps to identify interdependencies;
- Promotes integration of lessons learned and adoption of a forward-looking approach; and
- Supports consistent approach and enables cooperation.

10. *How does All-Hazards Risk Assessment relate to EM planning?*
    - All-Hazards Risk Assessment informs all pillars of EM planning.
    - EM planning based on an all-hazards approach recognizes that the causes of emergencies can vary greatly, but many of the effects do not.
    - EM Planning based on an all-hazards approach allows planners to address emergency functions common to all hazards in the basic plan instead of having unique plans for every type of hazard.
    - All-hazards EM planning supports the identification of common tasks and who is responsible for accomplishing those tasks.
11. *What EM training is available and where can it be provided?*
    - The Emergency Management Institute (EMI) and FEMA offers multiple courses available to all levels of employees, which are free online or traditional classroom courses.
12. *What tools are available to assist me through the EM planning process?*

In addition to offering several EM-related courses, EMI and FEMA provide you with guidance documents such as the EM Planning Guide and the Emergency Management Leading Practices.

- Moreover, workshops are also held in support of implementation of those EM leading practices.

- Finally, criteria for self-assessment and evaluation of Strategic EM Plans are established by NIMS, NRP, and other documents.

13. *How frequently will the EM Planning Guide be updated?*
    - The EM Planning Guide will be needs to be updated to reflect the integration of the All-Hazards Risk Assessment Framework.
    - Subsequent revisions will be made on an annual basis, or as the situation dictates and amendments will be made at that time.

References

Abrahms, Max. *What Terrorists Really Want: Terrorist Motives and Counterterrorism Strategy International Security*. Cambridge: MIT Press, 2008.

Alexander, David. *Confronting Catastrophe: New Perspectives on Natural Disasters*. Harpenden: Terra Publishing, 2014.

Allison, Graham. *A Failure to Imagine the Worst*. Cambridge: Belfer Center for Science and International Affairs, 2010.

Ansarian, Allamah Husayn. *The Islamic Family Structure* (4th ed.). Qom: Ansariyan Publications, 2010.

AOL Inc. Cyber-HUMINT. http://www.wow.com/wiki/Cyber-HUMINT. May 5, 2017.

Ast, Scott Alan. *Managing Security Overseas*. Boca Raton, FL: CRC Press, 2010.

Aujero, Annjel. Financial Risk Management. https://prezi.com/t9iypzfs9wna/financial-risk-management/. June 25, 2014.

Baird, Malcolm E. *"Phases" of Emergency Management*. Vanderbilt: Vanderbilt Center for Transportation, 2010.

Bank of England. *CBEST Intelligence-LED Assessment Guide*. Mountian View: Creative Commons, 2016.

Bankoff, Greg. *Mapping Vulnerability: Disasters, Development and People*. London: Earth Scan, 2004.

Barlow, John Perry. "A Declaration of the Independence of Cyberspace." *Electronic Frontier Foundation* (1996).

Bigley, Gregory, and Karlene Roberts. "The Incident Command System: High-Reliability Organizing for Complex and Volatile Task Environments." *The Academy of Management Journal* (2001): 44 (6): 1281–1299.

Bockstette, Carsten. *Jihadist Terrorist Use of Strategic Communication Management Techniques*. Garmisch-Partenkirchen: George C. Marshall Center Occasional Paper Series, 2008.

Boin, Arjen, and Mark Rhinard. *Managing Transboundary Crises: What Role for the European Union?* International Studies Review. 10. Oxford: Oxford University Press, 2008.

Buchanan, Sally, Paul Banks, and Roberta Pilette. *Emergency Preparedness Preservation Issues and Planning*. Chicago, IL: American Library Association, 2000.

Burton, Christopher G. "Environmental Risk and Hazards." *International Encyclopedia of the Social & Behavioral Sciences* (2001): 4655–4659.

Butler Review. *Review of Intelligence on Weapons of Mass Destruction*. London: The National Archives of the United Kingdom, 2006.

Committee on Judiciary and the United States Senate. *Federal Technology Service General Services Administration Before The Subcommittee On Terrorism Technology And Government Information*. Committee on Judiciary and the United States Senate. Washington, DC: Office of Information Assurance And Critical Infrastructure Protection Federal Technology Service General Services Administration, 2001.

Consultant.com. ISO 27001 Information Security Management System (ISMS). http://www.iso-consultants.com/iso-27001.htm. January 3, 2017.

Cooper, H.H.A. Experts Say the Terrorism Has No Religion. http://theislamophobia.blogspot.com/2017/08/Experts-say-The-terrorism-has-no-religion.html. January 6, 2017.

CPNI. Center for the Protection of National Infrastructure. https://www.cpni.gov.uk/. January 8, 2017.

Crockford, Neil. *An Introduction to Risk Management* (2nd ed.). Cambridge: Woodhead-Faulkner, 1986.

Department of Defense. *Emergency War Surgery 3rd Editon "Clinical Practice Guidelines."* Washington, DC: Department of Defense, 2004.

Department of Defense. *Sustaining U.S. Global Leadership: Priorities for the 21st Century Defense*. Washington, DC: Department of Defense, 2012.

Department of Defense. *Text Book of Military Medicine*. Washington, DC: Office of The Surgeon General at TMM Publishing, 2007.

Department of Homeland Security. *Capstone Document: Mass Fatality Management for Incidents Involving Weapons of Mass Destruction*. Washington, DC: Department of Homeland Security, 2005.

Department of Homeland Security. *National Incident Management System*. Washington, DC: Department of Homeland Security, 2008.

Department of Homeland Security. *National Strategy to Secure Cyberspace*. Washington, DC: Department of Homeland Security, 2014.

Department of the Army. *TM 3-11.42 Multi-Service Tatctics, Techniques, and Procedures for Installation Emergency Management*. Washington, DC: US Government, 2014.

DHHS (NIOSH). "Occupational Safety and Health Guidance Manual for Hazardous Waste Site Activities." *NIOSH* 22 February 2011: 85–115.

Drabek, Thomas. *Emergency Management: Principles and Practice for Local Government*. Washington: International City Management Association, 1991.

Faulstich, Gloser S. *Criticality Matrix*. Amsterdam: Elsevier, 2015.

Federal Aviation Administration. System Safety Process. https://www.faasafety.gov/gslac/ALC/libview_normal.aspx?id=6877. May 1, 2017.

FEMA. *Principles of Emergency Management Supplement*. Washington, DC: FEMA, 2007.

FEMA. *Heart and Hazard Identification and Risk Assessment Guide*. Washington, DC: FEMA, 2013.

Fennelly, Lawrence. *Crime Prevention Through Environmental Design* (3rd ed.). Oxford: Butterworth-Heinemann, 2013.

Fischer, Robert P., and Edward Halibozek. *Introduction to Security* (8th ed.). Burlington: Butterworth and Heinemann, 2008.

Forensic Science International. "Health & Safety Executive: Methods of decontamination." *Forensic Science International* (2013): 94: 61–71.

Garigue, Robert, and Andrew Mackie. "Provincial Action to National Security: A National Information Protection Agenda for Securing Government in Cyberspace." Ed. Information Protection and Assurance. CIO Conference, 1999.

Géron, Aurélien. *Hands-On Machine Learning with Scikit-Learn and TensorFlow: Concepts, Tools, and Techniques to Build Intelligent Systems* (1st ed.). North Sebastopol: O'Rielly Media Inc., 2017.

Graham, Mark. "Geography/Internet: Ethereal Alternate Dimensions of Cyberspace or Grounded Augmented Realities?" *The Geographical Journal* (n.d.): 179 (2): 177–188.

Greene, R.W. *Confronting Catastrophe*. Redlands, CA: ESRI, 2003.

Griffin, Roger. *Principles of Hazardous Materials Management*. Boca Raton, FL: CRC Press, 2009.

Groh, Maximilian. "Strategic Management in Times of Crisis." *American Journal of Economics and Business Administration* (2014): 6: 1–49.

Hamdan, Maha. *Motiviation of Terrorists*. https://www.linkedin.com/pulse/motiviation-terrorists-maha-hamdan. Janaury 3, 2015.

Harigel, Gert. *Depleted Uranium Weapons—A Threat to Human Health?* Santa Barbara: Nuclear Age Peace Foundation, 2017.

Harvard Kennedy School of Government. *Nuclear Terrorism FAQ*. Cambridge: The Nuclear Threat Initiative and Project on Managing the Atom, Belfer Center, 2007.

Hayden, M. Terrorism. Everipedia Plus. https://everipedia.org/wiki/terrorism/. February 4, 2015.
Hewitt, Ken, and Michael L. Sheehan. "A Pilot Survey of Global Natural Disasters the Past Twenty Years." *Natural Hazards Research Working Paper*, 1969.
Hillson, David, and Peter Simon. *Practical Risk Management: The ATOM Methodology*. Tysons Corner, VA: Management Concepts, 2012.
Hoffman, Bruce. *Inside Terrorism*. New York: Columbia Press, 2006.
Hubbard, Douglas. *The Failure of Risk Management: Why It's Broken and How to Fix It*. New York: John Wiley & Sons, 2009.
IAEM. *Certification Program Details*. Retrieved from International Association of Emergency Managers (IAEM): https://www.iaem.com/page.cfm?p=certification/getting-started. January 6, 2017.
IAEM. *International Association of Emergency Managers (IAEM)*. http://www.iaem.com/page.cfm?p=about/intro. August 22, 2016.
ICPEM. *Institute of Civil Protection & Emergency Management|Welcome*. March 4, 2015. London: Institute of Civil Protection and Emergency Management, 2015.
IMSM. *Cyber Warfare Threat Warning: Information & Data Security*. Wiltshire: IMSM, 2010.
Information Security Forum. The ISF is the world's leading authority on cyber, information security and risk management. https://www.securityforum.org/. 2017.
Integrated Security Committee. *The Risk Management Process for Federal Facilities: An Interagency Security Committee Standard*. Washington, DC: US Department of Homeland Security, 2016.
International Atomic Energy Agency. Incident and Trafficking Database (ITDB). http://www-ns.iaea.org/security/itdb.as. February 10, 2017.
James, Edward, and Erika Hayes James. "Linking Crisis Management and Leadership Competencies: The Role of Human Resource Development." *Advances in Developing Human Resources* (2008).
James, Erika Hayes. *Crisis Leadership*. SSRN. https://papers.ssrn.com/sol3/papers.cfm?abstract_id=1281843&rec=1&srcabs=224055. October 21, 2008.
Jones, David. *Nomenclature for hazard and risk assessment in the process industries*. Warwickshire, England: Institution of Chemical Engineers, 1992.
Koepsel, David. *The Ontology of Cyberspace*. Chicago, IL: Open Court, 2000.
Laqueur, Walter. *A History of Terrorism* (1st ed.). New York: Routledge, 2001.
Majors, Crystal. "The Future of Emergency Management." 2008.
Marin Operational Area Mass Fatality Plan MARIN OA EOP ANNEX. Marin: Marin County, 2012.
Mattox, Kenneth. *Trauma* (7th ed.). New York: McGraw-Hill Education, 2013.
McElreath, David et al. *Foundations of Emergency Management*. Chicago, IL: Kendall-Hunt Publishing Company, 2014.
McGraw Hill. *Water Decontamination*. New York: McGraw Hill, 2004.
Mckenna, Chris. *History of Terrorism*. New York: Columbia Press, 2012.
Mendell, Ronald. *The Quiet Threat*. Springfield, MA: Charles C Thomas Publisher, LTD, 2011.
MidEastweb. *The Security Council passed Resolution 687*. Retrieved from MidEastweb: http://www.mideastweb.org/687.htm. April 3, 1993.
Mission Areas, FEMA.gov, http://www.fema.gov/mission-areas; Core Capabilities. This page contains the 32 core capabilities identified in the National Preparedness Goal and is intended to assist everyone who has a role in achieving all of the elements in the Goal, 2017.
Mistovich, Joseph J., Keith J. Karren, and Brent Hafen. *Prehospital Emergency Care*. Upper Saddle River: Prentice Hall, 2013.
Moeller, Susan D. *Media Coverage of Weapons of Mass Destruction*. College Park: Center for International and Security Studies, 1998.

Morgan, O.W., Sribanditmongkol, P., Perera, C., Sulasmi, Y., and Van Alphen, D. *Mass Fatality Management Following the South Asian Tsunami Disaster: Case Studies in Thailand, Indonesia, and Sri Lanka*. San Francisco, CA: PLOS, 2006.

Morgan, O.W., Sribanditmongkol, P., Perera, C., Sulasmi, Y., and Van Alphen, D. *Mass Fatality Management Following the South Asian Tsunami Disaster*. PLoS Med, 2006.

Mughal, Yakub. *A Worthy Successor to the Quaid*. Karachi: Royal Book Company, 2012.

NACCHO. Managing Mass Fatalities: A Toolkit for Planning. National Association of County & City Health Officials: http://archived.naccho.org/toolbox/tool.cfm?id=1595 %20. December 5, 2016.

NACCHO. Toolbox a Free Collection of Local Public Health Tools. Retrieved from NACCHO: http://toolbox.naccho.org/pages/index.html. January 5, 2017.

Nasheri, Hedieh. *Economic Espionage and Industrial Spying*. New York: Cambridge University Press, 2005.

National Association of Medical Examiners. *Mass Fatality Plan*. Bethesda, MD: National Association of Medical Examiners, 2005.

National Disaster Medical System. *DMORT National Disaster Medical System*. Washington, DC: U.S. Department of Health and Human Services, 2016.

National Incident Management. *National Incident System Management*. Washington, DC: FEMA, 2008.

National Security Council. *The Administration • National Security Council*. Washington, DC: National Security Council, 2016.

Nemth, Charles. *Homeland Security Introduction to Principles and Practices* (2nd ed.). Baton Rouge, LA: CRC Press, 2013.

New Zealand Department of Internal Affairs. *National Civil Defence Emergency Management Strategy*. Wellington, New Zealand: New Zealand Department of Internal Affairs, 2008.

Niska, Richard W., and Iris M. Shimizu. *Hospital Preparedness for Emergency Response: United States*. Washington, DC: National Health Statistics Reports, 2008.

Occupational Safety and Health Administration. "Hazardous Waste Operations and Emergency Response (HAZWOPER)." *OSHA* (2006), 29 CFR 1910.120.

OSR Journal of Business and Management. "Risk Management—An Analytical Study." *OSR Journal of Business and Management* (2014): 8.

preventwmd.org. Prevention of Weapons of Mass Destruction Proliferation and Terrorism. http://preventwmd.org/. January 22, 2017.

QuickSeries. *Incident Command System (ICS)*. Jefferson City, MO: QuickSeries, 2008.

Ralph, Tom. *2016 Disaster Resource Guide*." Internal Revenue Service, October 11, 2016.

Renfroe, Nancy A. *Threat/Vulnerability Assessments and Risk Analysis*. Washington DC: National Institute of Building Sciences, 2016.

Renfroe, Nancy A., and Joseph L. Smith, PSP. *Threat/Vulnerability Assessments and Risk Analysis*. Washington, DC: National Institute of Building Sciences, 2014.

Revolvy. Chinese Information Operations. Chinese Information Operations and Information Warfare: http://chinawatchcanada.blogspot.com/2017/08/chinese-intelligence-worldwide -remains.html. 2017.

Reynolds, Lori E. Cyberspace. Retrieved from Marine Corps Cyberspace Command: https:// www.revolvy.com/topic/Cyberspace&item_type=topic. November 5, 2016.

Rodin, David. *War, Torture and Terrorism: Ethics and War in the 21st Century* (1st ed.). Hoboken: Blackwell Publishing, 2007.

Ronfeldt, Arquilla. "Cyberwar Is Coming." *Journal of Comparative Strategy* (n.d.): 12.

Ropeik, David. *Risk*. New York: Houghton Mifflin Company, 2002.

Royal Cornwall Hospitals. *Royal Cornwall Hospitals NHS Trust: Decontamination*. London: Royal Cornwall Hospitals, 2012.

Rudolph, Barbara. "Coping With Catastrophe." *Time*. February 24, 1986.

Sanders, Mick J. et al. *Mosby's Paramedic Textbook*. Sudbury, MA: Jones & Bartlett Publishers, 2011.
Sandman, Peter M. The Role of Apologizing in Crisis Situations. http://www.psandman.com/articles/busters.htm. June 28, 2005.
Schmid, Alex. *Political Terrorism: A Research Guide to Concept, Theories, Data Bases and Literature*. London: Elsevier Science Ltd, 1983.
Sensagent Corporation. The Alexander Litvinenko Poisoning. Nuclear terrorism. http://dictionary.sensagent.com/Nuclear%20terrorism/en-en/. November 20, 2016.
Sensagent. Cyberwarfare. http://dictionary.sensagent.com/Cyberwarfare/en-en/. October 20, 2016.
SIDS of the Pacific. The Environmental Vulnerability Index. http://www.vulnerabilityindex.net/. March 10, 2015.
Smith, Keith. *Environmental Hazards: Assessing Risk and Reducing Disaster*. New York: Routledge, 2001.
Soble, Jonathan. *Fukushima Keeps Fighting Radioactive Tide 5 Years After Disaster*. New York: New York Times, 2016.
Strate, Lance. "The varieties of cyberspace: Problems in definition and delimitation." *Western Journal of Communication* (1999): 382–383.
Strumfels, David. *Weapon of Mass Destruction*. New York: Science/Computing Gourmet, 2015.
Talbot, Julian, and Miles Jakeman. *Security Risk Management Body of Knowledge*. Hoboken, NJ: John Wiley & Sons, 2009.
Taylor, Craig, and Erik VanMarcke. *Acceptable Risk Processes: Lifelines and Natural Hazards*. Reston, VA: American Society of Civil Engineers, 2002.
Taylor, Penny. "Transporting and Disposing of Dangerous Goods in the US: What You Need to Know." National Chemical Emergency Centre, 2015.
The Committee of Sponsoring Organizations of the Treadway Commission. *Enterprise Risk Management — Integrated Framework*. Retrieved from The Committee of Sponsoring Organizations of the Treadway Commission: https://www.coso.org/Pages/erm-integrated framework.aspx. January 10, 2017.
The National Strategy to Secure Cyberspace. Washington, DC: Office of Homeland Security, 2003.
The Open Group. *Risk Taxonomy*. Berkshire: The Open Group, 2009.
Transportation Security Administration. Surface Transportation Vulnerability Assessments and Security Plans (VASP). Federal Register. https://www.federalregister.gov/documents/2016/12/16/2016-28300/surface-transportation-vulnerability-assessments-and-security-plans-vasp. December 16, 2016.
Traynor, Ian. *Angela Merkel: NSA Spying on Allies Is Not On*. New York: The Guardian, 2013.
Trunkey, Donald. *Current Therapy of Trauma and Surgical Critical Care* (1st ed.). Philadelphia: Mosby, 2008.
UNDRO. Case Study: Town Relocation, disasterassessment.org/documents/. 1982.
United Nations. *50/53. Measures to Eliminate International Terrorism*. January 29, 1996. Resolution adopted by the General Assembly. http://www.un.org/documents/ga/res/50/ares50-53.htm.
United Nations. *Small Arms Conference Urged to Forge Comprehensive Action Plan*. New York: United Nations Conference on the Illicit Trade in Small Arms, 2001.
U.S. Army. *FM 3-11.4 Multiservice Tactics, Techniques, and Procedures for Nuclear, Biological, and Chemical (NBC) Protection*. Washington, DC: U.S. Army, 2003.
U.S. Army. *FM 3-11.9 Potential Military Chemical/Biological Agents and Compounds*. Washington, DC: U.S. Army, 2005.
U.S. Army. *Defense Support of Civil Authorities*. Washington, DC: U.S. Army Training and Doctrine Command, 2005.

U.S. Army. *FM 3-11.3 Multiservice Tactics, Techniques, and Procedures for Chemical, Biological, Radiological, and Nuclear Contamination Avoidance*. Washington, DC: U.S. Army, 2006.

U.S. Army. *FM 3-11.5 Multiservice Tactics, Techniques, and Procedures for Chemical, Biological, Radiological, and Nuclear Decontamination*. Washington, DC: U.S. Army, 2006.

U.S. Army. Composite Risk Management Process. http://www.wv.ngb.army.mil/campdawson/documents/risk%20management%20process.pdf. January 2010.

U.S. Army. *FM 3-11 Multi-Service Doctrine for Chemical, Biological, Radiological, and Nuclear OperatioNS*. Washington, DC: U.S. Army, 2011.

U.S. Army. *Emergency War Surgery*. Washington, DC: U.S. Governement, 2016. <http://www.cs.amedd.army.mil/borden/FileDownloadpublic.aspx?docid=68aca9a0-9cd7-4d8f-a17f-a4c01264daef>.

U.S. Commission on National Security/21st Century. *Road Map for National Security*. Washington: U.S. Air Force, 2001.

U.S. Department of Defense. *Military Support to Civil Authorities (MSCA) Department of Defense Directive 3025.01. J*. Washington, DC: U.S. Department of Defense, 1993.

U.S. Department of Defense. *Strategy for Homeland Defense and Civil Support*. Washington, DC: U.S. Department of Defense, 2005.

U.S. Department of Defense. *Defense Support of Civil Authorities Policy of the Twenty-First Century*. Washington, DC: U.S. Department of Defense, 2006.

U.S. Department of Defense. *Assistant Secretary of Defense for Homeland Defense and Americas' Security Affairs (ASD(HD&ASA)). Department of Defense Directive 5111.13*. Washington, DC: U.S. Department of Defense, 2009.

U.S. Department of Energy. *Vulnerability Assessment Methodology, Electric Power Infrastructure*. Washington, DC: U.S. Department of Energy, 2002.

U.S. Department of Health and Human Services. *Disaster Mortuary Operational Response Teams (DMORTs)*. Washington, DC: U.S. Department of Health and Human Services, 2015.

U.S. Department of Homeland Security. *National Response Plan*. Washington, DC: U.S. Department of Homeland Security, 2004.

U.S. Department of Transportation. *Emergency Response Guide 2016*. Washington, DC: U.S. Department of Transportation, 2016.

U.S. Government Publishing Office. *22 U.S.C. 2656f - Annual Country Reports on Terrorism*. Retrieved from U.S. Government Publishing Office: https://www.gpo.gov/fdsys/granule/USCODE-2010-title22/USCODE-2010-title22-chap38-sec2656f. January 15, 2017.

U.S. Government. *The Classified National Security Information Protection Act*. Washington, DC: GPO, 2003.

Walzer, Michael. A discussion with Michael Malzer. Terrorisme.net. 2006.

Wisner, Ben. *At Risk: Natural Hazards, People's Vulnerability and Disasters* (2nd ed.). New York: Routledge, 2003.

Wormuth, Christine. *The Future of the National Guard and Reserves: The Beyond Goldwater-Nichols Phase III*. Washington, DC: Center for Strategic and International Studies, 2006.

Wright, Joe, and Jim Harmening. *Computer and Information Security Handbook*. Burlington: Morgan Kaufmann Publications, 2009.

Zetter, Kim. *An Unprecedented Look at Stuxnet, the World's First Digital Weapon*. New York: Wired, 2014.

Index

D

E

F

Q

R

S

T